U0168894

以模型思考建筑

建筑中的抽象与几何

[芬] 海基宁－科莫宁建筑事务所 著 | 景璟 李桂春 译

中国电力出版社
CHINA ELECTRIC POWER PRESS

图书在版编目（CIP）数据

以模型思考建筑：建筑中的抽象与几何：汉英对照/（芬）海基宁－科莫宁建筑事务所著；景璟，李桂春译．—北京：中国电力出版社，2021.7

（当代设计新思维论丛／方海主编）

ISBN 978-7-5198-4706-7

Ⅰ.①以… Ⅱ.①芬…②景… ③李… Ⅲ.①模型（建筑）－设计－汉、英 Ⅳ.①TU205

中国版本图书馆CIP数据核字（2020）第099845号

出版发行：中国电力出版社
地　　址：北京市东城区北京站西街19号（邮政编码 100005）
网　　址：http://www.cepp.sgcc.com.cn
责任编辑：王　倩（010-63412607）
责任校对：黄　蓓　郝军燕
责任印制：杨晓东
装帧设计：锋尚设计

印　　刷：北京瑞禾彩色印刷有限公司
版　　次：2021年7月第1版
印　　次：2021年7月北京第1次印刷
开　　本：710毫米×980毫米　16开本
印　　张：10.75
字　　数：172千字
定　　价：88.00元

目　录
contents

序

建筑师的模型思维
从“抽象与几何”到建筑杰作

方海

海基宁-科莫宁建筑事务所是当代最活跃也是最有代表性的芬兰建筑师团队之一。两位创始人兼主要合伙人米高·海基宁和马可·科莫宁都曾长期兼任阿尔托大学（前赫尔辛基理工大学）建筑学院的教授，是芬兰建筑界学者型建筑师的典型代表，也是展现芬兰建筑教育的一个成功元素。在他们的建筑事务所中，绝大多数员工都曾是他们的学生，其中的许多人从事务所离开后建立了新的设计工作室，甚至比他们的事务所规模更大，这让人强烈感受到了芬兰建筑生生不息的传承与活力。

与此同时，每年的寒暑假期间，会有世界各地建筑或设计专业的师生前往海基宁-科莫宁建筑事务所参观和交流。2018年夏天，笔者带领一组来自中国的建筑学专业师生前往事务所参观，受到海基宁教授的热情接待。同学们被办公室中随处可见的建筑模型所吸引，这些模型由不同材料制作而成，它们比例不同、风格各异、创意非凡、引人入胜，构成一个活脱脱的小型建筑博物馆……当同学们走到海基宁教授的办公桌前时，惊讶和喜悦的心情随即达到高潮，因为同学们看到海基宁教授的办公桌或绘图桌上不仅堆满了由不同材料制成的、不同设计阶段的建筑模型，而且遍布着制作模型的各类基本工具，于是有了如下的对话。

学生：海基宁教授，您平时的工作都是做模型吗？

教授：对我来说，在项目的早期阶段，用模型来做草图是必不可少的。用描图纸、硬纸板和透明胶片等桌子上能找到的材料来塑造简单的构图，是我思考过程的一部分。即便一张纸，将其折叠，也可能传达出解决一个方案的基本思想。用纸板和胶水表达思想，有时比用3D屏幕更快。

学生：那么，您办公室中这么多模型都是您和您的同事亲手制作的吗？

教授：并不是，你在我们办公室看到的大部分模型都是由专业模型公司制作的。这些模型是在实际设计过程结束后做出的，甚至是在建筑或项目需要实现时才做。它们用于向客户展示，或是作为竞赛项目的一部分，或是展览的需要。我们没有制造出那种完美模型的专业技术或机器。我们在设计之初做的很多草图模型，都在帮助我们推敲设计后，被扔进了工作台下的垃圾桶。

学生：您办公室的上百件建筑模型仿佛使用了我们能想象到的所有材料，您对做模型的材料有规定吗？比如，您在教学中是如何要求学生做模型的？对学生使用的材料有哪些限定和建议？

教授：手边的任何材料都可以用来做模型。手绘的三维草图应尽可能画得简单与快速。你的想法越详细，做出的模型就越清晰；你也可以用不同材料更好地表达你的设计。在开始阶段，研究物体的形式或许就足够，但之后你可以将重点放在体块的透明度和实体部分的表达等方面。在这个过程中，细节并不重要，应在整体方法上多下工夫。在数字世界里，学生们很容易迷失，所以我们试着让学生们在接触数字世界之前就开始玩模型。我想这很健康，因为不用盯着电脑屏幕，而是用手来做。

1．芬兰建筑师的“抽象与几何”在中国的遭遇

中国的城市在过去三十年拥有了全球最大的建筑建造数量，世界各地的建筑师涌入中国参与设计和建造，他们和数量庞大的中国建筑师一道，将中国大批城市大幅度改观，用令人难以置信的建造速度，推动着中国的城市化进程。总的来说，欧、美、日多国的建筑大师们不断为中国建筑界带来了最时尚的建筑理念和最先进的建筑技术与材料，同时也带动了中国建筑界同仁在合作中的共同成长。在一段时间中，盘点各大城市最优秀、最引人注目的建筑，大多都是出自欧、美、日国家和中国香港与台湾地区的建筑师。中国本土建筑师“主场缺席”的主要原因是什么？有多方面的原因。

芬兰建筑大师拜卡·萨米宁是最早进入中国的北欧建筑师之一，也是在中国获得成功的西方建筑大师之一。在过去的二十年间，萨米宁建筑事务所在中国参加过百余次国际建筑设计竞赛，并二十余次获得第一名，已完成大量的建筑项目，其中无锡大剧院、成都云端大厦、福州海峡文化艺术中心与大型公共建筑都成为业内标杆，受到业主和国内外建筑同仁的一致好评。近年来，每当他与我聊起中国当代建筑和他在中国的设计实践，已八十多岁高龄的萨米宁教授都会感慨颇多。他会谈起中国传统文化的博大精深，也会提到中国业主的“高深莫测”；他时常羡慕中国古代的设计智慧，也会感叹当代中国“快速发展”给传统文化带来的冲击。对中国当代建筑师同仁，萨米宁教授称赞其广收博取的学习能力，但同时也善意指出其设计观念上的偏差，尤其是抽象能力的不足。实际上，“抽象思维能力不足”的状况并非限于中国建筑师群体，很多人都普遍习惯于具象思维，看看中国各地的城市雕塑，“栩栩如生”往往是最高赞誉。萨米宁教授坦言，“抽象思维能力不足”从根本上制约着中国建筑师潜在聪明才智的发挥，而对几何、尺度、比例诸多方面的漠视或不够重视，更造成中国建筑师总体设计上的单调，并因为这种“单调”而丧失“趣味”，从而导致总体建筑质量的缺失。

萨米宁教授坦言，也许是中国历史悠久的具象文化在广大人民心中烙下极深的印象，具象思维总是轻易凌驾于抽象观念之上。萨米宁热爱中国文化，并时常在设计中融入中国文化中的诸多设计因子。这些中国特有的设计因子在芬兰建筑师脑中会被过滤成一系列新型创意符号，并与西方文化传统中源远流长的几何学设计手法融会贯通。于是，在萨米宁的创作中，武汉湖滨酒店“平沙落雁”的身影，武汉国际机场“雄鹰展翅”的气势，无锡大剧院“蜻蜓点水”的曼妙，福州海峡文化艺术中心“风帆竞渡”的豪迈，都立刻赢得业主的赞誉，其中部分是“芬兰式象征设计”手法演化为中国城市的地标形象。然而，对现代城市建筑而言，显而易见的象征并非总是最佳选择，在大多数情形下甚至并不得当。人类科技的高度发展源自理性的抽象思维，与高科技和新材料息息相关的现代建筑必然要以抽象思维为主导，这种抽象思维的主导作用不仅表现在高层建筑和大跨度建筑上，也会直接用于高科技和高强度材料支撑的建筑类型上，并同样应体现在中小型文化建筑和景观设计类型中。萨米宁教授曾多次对笔者讲述“芬兰建筑师的几何与抽象”在中国的境遇，其中有迷惑，也有无奈。有一次，萨米宁用快捷的草图为某甲方讲解当地城市展览馆屋顶结构的概念方案，可对方不停追问草图中那些连接中断的线条在实际建造中如何处理。萨米宁一方面解释“这只是构思草图，那些中断的线条会在之后阶段的设计中圆满填补好”，另一方面心中感叹这种想法其实是抽象思维方面的某种缺失。还有一次，萨米宁建筑事务所参加上海某社区学校的建筑设计竞赛，并进入最后的决赛论证，萨米宁教授的方案充分展示了芬兰当代学校建筑设计的实力，以精巧的几何构图贯穿于平面布局、立面造型和室内空间布置中。本已被评审专家确认为实施方案，然而因种种原因，最终专家们选择了一种“宁可放弃功能，也要充满象征”的方案，让萨米宁教授直呼无奈。这样的经历，笔者认识的许多欧美建筑师在中国都曾遇到过。这可以看作是中国现代建筑历程的必经之路，过于快速的发展必然会带来诸多的矛盾和困境。

令人欣慰的是，近年来互联网的高速发展，使得建筑批评进入千家万户。各种机构不仅时常报道出中国大地上出现的“小白宫”和“明清商业街”，而且每年会评选出“中国十大年度最丑建筑”，这些可看出中国全民审美情趣的转变与提升。当广大群众传看“福禄寿酒店”“金元宝大厦”“大铜钱拱门”，以及“大茶壶”和“大铁锅”的图片时，已经明显感受到在潜移默化中，人们从“象征的图像”转向“抽象的思考”。而中国广大建筑师也会通过抽象思维训练，注重从几何学入手，逐步进入以“比例和尺度”为核心的理性建筑。

2. 两本书：《古典建筑的起源》（*Origins of Classical Architecture*）和《建筑理论》

耶鲁大学出版社2014年出版了著名学者马克·威尔逊·琼斯（Mark Wilson Jones）所著的新作《古典建筑的起源：古希腊的庙宇、秩序及给神的礼物》（*Origins of Classical Architecture: Temples, Orders and Gifts to the Gods in Ancient Greece*）。该书试图从建筑师的视角探寻西方世界延续至今的古典建筑传统中最具代表性的内容，即以多立克、爱奥尼和科林斯为代表的古典柱式。古希腊建筑前承古埃及建筑，后启古罗马建筑，而后与古罗马建筑一道开启意大利文艺复兴，继而引发全欧洲文艺复兴，最后进入现代并引领全球文明发展。从古埃及、古西亚到古希腊、古罗马，其古典建筑中最核心的构成元素即柱式。它们是所有古典建筑中结构体系的灵魂，其形态直接源自大自然，尤其是森林树木，即大树的抽象与简化形态。抽象后的大树在古埃及与古西亚漫长的建筑演化中经历代建筑师的几何规划，最终在古希腊时代形成，而至今仍然可看到的大量以多立克、爱奥尼和科林斯为代表的西方古典柱式实例，随后被古罗马全盘收入古罗马建筑体系，最终成为欧美古典建筑体系的核心元素，由此使西方建筑传统牢牢建立在“抽象与几何”的思维基础之上。这样的理性思维与大机器时代的工业化和现代化同步吻合、携手并进，带动并促成现代建筑的高歌猛进。

琼斯教授在长期的考察和研究中发现古希腊柱式的成熟与定型，是在相对而言非常短暂的一个时期完成的，并与当时文化的发展、国际交流和艺术创新的渴望密切相关。流传至今的古希腊建筑大多数都是神庙，它们毫无疑问也是当时希腊社会中最重要的建筑。古希腊是一个人神共生的社会，由此形成众神既有高高在上的崇高一面，同时又与人类社会息息相关，古希腊神庙因此取得了一种独一无二的神圣地位。神庙是民众供奉众神的场所，其建筑本身也是供奉给神的重要礼物，因此尽显豪华与崇高。尤其为保存天长地久，必用最好的石材、高大的尺度、庄重的柱式和豪华的构造形成神庙的基本面貌，创造出西方古典世界一种全新的建筑形式。这种充满崇高意味和宗教感的古典柱式及它们组合成的神庙形态，最终成为西方建筑延续不断的脉络灵魂，历经中世纪千年巍然不移，经文艺复兴又发扬光大并正式成为全欧洲的建筑遗产。其核心为“抽象与几何”的设计理念完全契合工业革命后的现代化大都市发展，同样它也是现代建筑与现代设计的核心理念。反观中国，我们祖先发明的木构建筑体系虽然同样精美，并曾在历朝历代让世界叹为观止，然而木构自身的弱点使之遇火即焚，难以持久。中国古代建筑中能与西方古典柱式相对应的当属斗拱，它们都是承上启下的结构元素，同时也是机巧繁复的智力游戏。随着时代的发展，斗拱的装饰功能逐渐大过结构上的含义。中国古代建筑中的斗拱，有些学者认为也是源自对大树树冠的仿效，但缺少西方古典柱式“抽象与几何”的思维基础，因此难以融入现代社会，这也是中国现代建筑师的背景软肋。从刘敦桢、梁思成、杨廷宝和童寯开始，一代又一代中国建筑师一直进行着前仆后继的探索，但大多都在民族式与国际式、城镇化与乡土风之间徘徊，很少有人真正关注建筑基本思维中的“抽象与几何”。而这种思维，每时每刻都蔓延在海基宁-科莫宁的办公室中，也真真切切地贯穿于芬兰和欧、美、日等国的优秀建筑中，值得中国建筑师深思、再深思。

2018年6月，北京出版集团出版了德国学者伯恩德·艾弗森主编、唐韵等译的《建筑理论：从文艺复兴至今》。该书初版时间为2003年，由德国著名的塔森（Taschen）出版社隆重推出，以八百余页的篇幅，详细介绍了意大利、英国、法国、德国、西班牙等国建筑精英对建筑及相关问题的论述，同时配以大量插图。在西方，建筑理论一般都追溯至古罗马建筑师和工程师维特鲁威的《建筑十书》，该书作为建筑学最古老的论著，在全世界以各种文字一版再版。维特鲁威的著作系统记载了古罗马时代的主要建筑类型，由此忠实传递着以古典柱式为核心的西方建筑抽象思维传统和几何表达手法。《建筑十书》不仅是有关西方古典建筑及其原则的珍贵信息，而且也是文艺复兴以来所有建筑思考和建筑理论著作的基石。

克里斯托弗·托奥斯在该书的“导读”中从“建筑与文学”的话题谈起，介绍了“人文主义”“文艺复兴”“宫廷文化与城市文化”“书籍”“书籍中的插图”“柱式”“建筑理论与乌托邦社会”“理论与反理论”等。就西方建筑自身的发展而言，柱式是最重要的“晴雨表”，一方面表示它成为西方建筑的基石元素，另一方面也代表着西方世界不同时代、不同国家和地区的建筑发展与演化，无论千变万化，以“柱式”为象征的“抽象与几何”的思维模式始终如一。从建筑理论的发展脉络看，欧洲各国的建筑发展在进入现代之前基本是同步的。在意大利，从阿尔伯蒂、维尼奥拉、帕拉第奥到皮拉内西，他们的建筑理论始终都在讨论古典建筑模式尤其是柱式的复兴与规范；在法国，从奥内库尔、佩罗、布朗德尔到勒杜克，他们用更多的实例推动柱式的发展；在美国，从舒特、钱伯斯、普金到拉斯金，他们重视测绘古建筑遗址，将古老的柱式在记录中进行再创造；在德国，从丢勒、格伦特、申克尔到森佩尔，他们用德国人的精准和铜版字的细致，不仅翔实记录古典世界中最重要的建筑实例，也深入探讨多种方式的透视表达技法。至于20世纪的建筑理论，一方面建立在西方古典传统的基础之上，另一方面又紧密结合新时代科技和新材料的蓬勃发展。从霍华德、路斯、陶特，到柯布西耶、格罗皮乌斯、赖特，再到吉迪翁、文丘里和库哈斯，他们热烈探讨着时代的热门建筑话题，并试图给出建议和答案。他们的研讨和探索从来没有离开过“抽象与几何”的理性思维。《建筑理论：从文艺复兴至今》一书的理论主题就是对古希腊柱式进行持续不断探讨，并在这种探讨中保持着以“抽象与几何”为核心元素的理性思维，这是现代建筑的重要基石，应该引起中国建筑界对中国传统建筑遗产的新一轮研讨。

3. 现代建筑大师的创作灵魂：抽象与几何

现代建筑运动中的每一位大师都以各自不同的设计理念和手法发展和丰富着现代建筑，他们共有的创作灵魂就是“抽象与几何”。它是理念也是手法，由此形成现代建筑发展的主旋律。从某种意义上讲，每一代现代建筑大师都以自己独特的方式解读抽象的含义，同时用几何的手法将现代科技和材料转化为时代前沿的空间。

赖特在75岁时完成的自传，自1932年出版后不断再版，直到1943年发行的最终定本，包括了全部五卷内容。该版本随后开始在全球以不同语言出版，上海人民出版社在2014年出版了杨鹏译的《一部自传：弗兰克·劳埃德·赖特》。该书五卷分别题为家族、入行、事业、自由和形式，赖特本人亲自为每一卷的插页设计了抽象线条图案，展示了他运用几何图案表达创作意境的娴熟手段。终其一生，无论是城镇规划还是建筑，无论是室内还是家具，无论是地毯设计还是日用器皿设计，赖特都始终如一地用几何学的理解创造着人类的居所，以及其周边环境的方方面面。2005年，英国著名的菲登（Phaidon）出版社推出美国华盛顿大学建筑学教授罗伯特·麦卡特（Robert McCarter）主编的新著《论弗兰克·劳埃德·赖特：建筑理论读本》（*On and By Frank Llond Wright: A Primer of Architectural Principles*），其中包括赖特本人所写的三篇文章，“建筑的起因（In the Cause of Architecture）”“平面的逻辑（The Logic of the Plan）”和“自然中的材料（In the Nature of Materials）”。它们直接表达了赖特自己的设计哲学，即从大自然中获得灵感和启发元素，而后用几何和抽象的思维手法表达出空间的逻辑性，再用生态的、合适的材料来展现空间的逻辑关系，继而创造出与大自然和谐共处的现代建筑杰作。该书中的其他文章来自肯尼斯·弗兰普顿（Kenneth Frampton）、科林·罗（Colin Rowe）等十四位学者，他们的论述及数百幅来自赖特档案馆的构思草稿和附图，非常清晰地说明了赖特如何深度研习几何学原理，并将之用于空间的创造与演化过程中。

作为现代建筑和国际式风格的发源地和大本营，包豪斯也是在现代建筑、设计和艺术诸领域全方位提倡和实践“抽象与几何”思维模式的典范。包豪斯的四位校长都是世界级建筑大师，都在城市规划、建筑和设计领域取得了划时代成就。包豪斯开创者兼首任校长格罗皮乌斯对建筑、设计和艺术及其与抽象观念的关系有极为深刻的理解。他的建筑作品石破天惊，使其立刻成为新时代精神的代表。第二任校长汉斯·迈耶全身心致力于将“抽象与几何”的思维观念整合到社会改革和城乡集合住宅建筑当中。第三任校长密斯率先用抽象的拼贴艺术观念来表达其非常前卫的流通空间观念和玻璃摩天楼构想，同时用极为严整的几何学手法构建其建筑细节，开创了现代大都市的建筑规范。包豪斯的第四位校长是世界级建筑大师布劳耶尔，他则在建筑和家具两个领域同时展现其超前的创新意识。他对抽象与几何模式的运用出神入化，开创了现代建筑的新型造型语言。包豪斯的建筑大师群体对“抽象与几何”思维模式的理解和运用是划时代的，也是水到渠成的结果，因为包豪斯正是现代抽象艺术的诞生地和重镇之一。现代抽象艺术最重要的开创者和顶级大师如蒙德里安、马列维奇·凡·杜斯堡、利西斯基等都曾在包豪斯授课；另外几位更重要的抽象艺术开创者如康定斯基、克利、莫霍利-纳吉、阿尔伯斯、费宁格、施莱默等则是包豪斯的全职教师。他们受格罗皮乌斯的感召来到包豪斯，以不同的切入点发现大自然的抽象奥秘，并用几何、色彩、材料、科技和影像构筑起抽象艺术的大厦，影响了全世界的艺术创意和建筑与设计思维。包豪斯虽然只存在了十九年，但它是现代艺术与建筑思想的不灭火种。包豪斯的大师和学生们，随后在欧洲和美国建立了乌尔姆设计学院、黑山艺术学院和芝加哥新包豪斯，从而使包豪斯的抽象思维模式和设计创意理念传遍世界。

与包豪斯的四位建筑大师同时开创现代建筑新纪元后，柯布西耶同时兼具艺术家和建筑师的身份，艺术的创意始终以一种内在的方式指引着他作为建筑师的工作。他与奥赞方共同创建的纯粹主义学派虽受主体派启发，却与三维空间融为一体，实际上是用“抽象与几何”的思维导向发展出的一种物像表达。柯布西耶建筑生涯的前期以萨沃伊别墅为代表，该作品是一种以“抽象化的欧式几何”为主导的空间组织和立面演化；其建筑生涯的中后期以朗香教堂为代表，该作品是一种以“北欧几何化的抽象模式”主导的多重跨界空间构建和立面的自创演化；其建筑生涯的晚期代表作印度昌迪加尔新城中心建筑群，它则是柯布西耶式抽象意境与几何尺度的完美结合。柯布西耶贯穿其一生的模度系统研究是他对现代建筑中“抽象与几何”思维模式的理论归纳和科学总结。其精细的模度研究使之对建筑空间、比例、尺度与人体的关系进行多层面、多向度的比较分析，从而使柯布西耶的建筑作品充满创意和启迪意味，经久耐看，发人深省，历久弥新。虽然作为现代绘画大师之一的柯布西耶在现代绘画和雕塑领域中都占有一席之地，但对于模度研究，柯布西耶则完全在用科学家的态度进行持久而深入的探索，力图达到现代科学的高度，并曾为此专程携带有关模度的书稿前往普林斯顿拜访当代科学大师爱因斯坦，受到爱因斯坦的鼓励和赞赏。柯布西耶对建筑、城市、日用设计和“抽象与几何”思维模式的科学研究始于1912年出版的《德国装饰艺术运动研究》（*A Study of the Decorative Art Movement in Germany*），并于1923年出版论文集《走向新建筑》，奠定了其作为现代建筑首席发言人的地位。而他对建筑的系统性科学研究持续终生，陆续出版的著作包括《一栋住宅，一座宫殿：建筑整体性研究》《现代建筑年鉴》《今日的装饰艺术》《人类三大聚居地规划》《明日之城市》和《精确性：建筑与城市规划状况报告》等。柯布西耶的研究性建筑实践为后世建筑师树立了伟大的榜样。随着现代建筑逐步进入多元化、全球化和信息化阶段，各地建筑师对建筑的理解方式和研究态度也呈现多元化，柯布西耶和包豪斯之后的建筑师一方面看到了榜样的光辉，另一方面也开始体会并摸索适合自己的创意途径。

2017年，江苏凤凰科学技术出版社出版了戴维·B. 布朗宁和戴维·G. 德龙著、马琴译的《路易斯·康：在建筑的王国中》。该书让我们看到了现代建筑表达的另一种途径。美国著名建筑史学家文森特·斯科利在该书的绪论中明确揭示了路易斯·康如何在多年的建筑史教学中发现自己的建筑主题。“康在晚年发现了如何把古罗马的废墟转变成现代建筑，这种关系表面上看似完全不可能，但是康在萨尔克生物研究所之后所有的建筑作品中，成功地实现了这种转变。”康在教授建筑史的岁月中多次去埃及、希腊和意大利参观古代建筑遗址，潜心体会古代建筑经典永恒的“抽象与几何”序列带给自己的心灵震撼，并在随后的大量建筑实践中将埃及金字塔、古希腊神庙和古罗马市场的柱式空间直接引入其现代建筑项目中。“最重要的是，路易斯·康在艾哈迈达巴德（印度管理学院）和达卡（国民议会大厦）中的‘砖的秩序’来自从皮拉内西的创造中提炼出来的罗马的砖和混凝土结构，而达卡诊所的门廊外形类似于勒杜克描绘建筑师全视野的画。”

贝聿铭早年师从于格罗皮乌斯和布劳耶尔，得以内窥现代建筑理念的精华，对“抽象与几何”有非常明确的理解。其早期作品清晰显示了布劳耶尔“几何式抽象混凝土”造型的痕迹，而其成名作华盛顿国家美术馆东馆则将布劳耶尔的设计风格发挥到极致。同时他开始建立自己对几何构图和抽象空间尺度的观念，并在之后的设计实践中不断加入对经典建筑遗址的现代理解，形成强烈而充满历史感和文化气息的设计风格。这种震惊世界的设计风格首先在巴黎卢浮宫亮相。卢浮宫改建项目举世瞩目，贝聿铭深知一般的设计主题和手法难以服众，于是采用建筑史上最具盛名的古埃及金字塔形式。与康对古埃及、古希腊、古罗马古典遗址在使用空间方面的直接借用不同，贝聿铭借用了金字塔的几何比例，然后用新时代最受宠爱的材料——钢和玻璃，做出金字塔系列，放在卢浮宫广场。此举立刻轰动全球，法国民众从痛恨、不解到最爱，从另一个角度反映出贝聿铭创作手法的大胆和高超。作为华裔美国建筑大师，贝聿铭早在中国改革开放之初即受邀设计了北京香山饭店，这使他很自然地成为第一位进入中国建筑设计市场的国际建筑大师。在香山饭店的设计中，贝聿铭将创意的思维转向中国传统建筑和园林，但都用现代主义的“抽象与几何”思维模式进行解读，于是我们得以看到既具有明显中国园林风貌又充满现代感的香山饭店。一代国际建筑大师贝聿铭晚年的设计手法炉火纯青，登峰造极，其娴熟的抽象理念和几何空间手法总能与世界各地的不同文化传统融会贯通，进而产生一系列贝式“抽象与几何”的现代建筑经典案例，如日本美秀美术馆、中国苏州博物馆新馆和卡塔尔多哈伊斯兰博物馆等。

当代建筑师从前辈大师的创意理念和设计实践中体会到了抽象理念和几何尺度的深意，于是开始挖掘自己的文化特色和专业背景，致力于创造出属于自己的建筑风格。瑞士建筑大师博塔自幼受古罗马、古希腊风格的熏陶，像路易斯·康一样，从古希腊柱式和古罗马拱券中获得灵感，再用砖石将它们进行现代化的几何排列，使之服务于现代化的功能。日本建筑大师安滕忠雄自学成才，在其成长期的欧洲建筑考察中深深迷恋于柯布西耶的模度研究，更被柯布西耶和北欧几位建筑大师的混凝土杰作彻底吸引，于是做出坚定且偏执的决定，即用几何抽象的空间塑造理念打造现代混凝土建筑，立志将混凝土这种材料用抽象与几何的思维模式研究到极致，创作到极致。已故的当代建筑女杰扎哈·哈迪德则从数学研究入手，最终跻身于现代建筑殿堂。哈迪德在大学时代已成为小有成就的青年数学家，但她对建筑和绘画情有独钟，于是进入英国AA学校学习建筑，从数学的诸多定律和理念入手发展出一种只有少数人能看懂的建筑绘画，其中的非欧几何元素预示着不久之后横空出世的一系列建筑杰作。在全世界的质疑和惊叹中，哈迪德借助现代计算机和工程学的发展，使之融入自己由非欧几里得几何入手的建筑方案中。她先用混凝土材料顺利建成了位于德国的维特拉消防站，从此之后其惊世的建筑创作才华无法遏制，辅以钢与玻璃材料的全方位使用。哈迪德在由数学所导入的建筑王国中游刃有余。另一位超凡脱俗的建筑奇才丹尼尔·里伯斯金则从不同的角度进入一种独一无二的绘画，而后将其导入建筑设计的实践。里伯斯金博览群书，尤其对建筑史和艺术史论情有独钟，后被老萨里宁创办的美国匡溪艺术学院聘为教师，在轻松愉悦的教学之余，他开始发展出一套基于伊斯兰细密点与文艺复兴建筑点的一种立体几何通景点，这种前无古人的独特绘画非常引人入胜但又难以言说。它们将几何与空间的关系在抽象与具象之间无止境地挥洒，从而引发观者无穷的想象力。如同人们对哈迪德建筑绘画的态度一样，几乎所有的业内人士都认为里伯斯金的几何通景点只是一种数学与绘图游戏，然而里伯斯金最终用事实证明了其过人的设计才华。当犹太人纪念博物馆最终建成，立刻成为柏林的新地标之一时，人们也开始接受对现代建筑的又一种不同解读。当今世界还有一位独步世界的建筑大师，秉承达·芬奇和高迪的天才，从大自然中获取对抽象与几何的全新理解，创造出介于雕塑和建筑之间的一系列艺术佳作，他就是西班牙当代设计大师卡拉特洛瓦。卡拉特洛瓦拥有建筑学和工程学的博士学位，是一位杰出的工程师、建筑师、艺术家、发明家和科学家。他以达·芬奇为榜样，对大自然充满好奇；他受前辈建筑大师高迪影响至深，对人体骨骼结构及其运作原理兴趣盎然；而他受到的工程学训练又使他能用充满逻辑的系统建立自己对抽象思维和几何尺度模式的理解，从而建立自己独立于世的建筑创作系统。他的建筑和桥梁作品，每一座都成为当地的景点和地标。

中国当代建筑师还在摸索中，改革开放让我们能够全方位地接触上述每一位建筑大师的思想和作品，而众多巨大的建筑项目更为中国建筑师提供了无止境的发挥创意舞台。中国建筑师还缺什么呢？我们当然期待更好的业主、更合理的造建制度、更成熟的设计教育体系等，然而，我们最需要做的其实是沉下心来仔细且系统地思考自己的设计方法和创作理念，力争从根本上建立自己对“抽象与几何”思维模式的真正理解，并在这种“建立思维模式”的过程中精心审视本民族的文化背景、民族符号和传统设计智慧，同时像前辈大师一样，热爱大自然，观察大自然，将人类设计遗产视为自己的灵感宝库，为此，我们有足够的期待。

4. 芬兰建筑学派的形成

老萨里宁是芬兰建筑学派的开山鼻祖，是北欧设计学派的奠基人，也是19世纪末至20世纪初芬兰民族浪漫主义建筑运动的旗手。尽管他的后半生活动大都在美国，是美国现代建筑顶级大师之一，同时也是美国现代设计和艺术教育的重要奠基者之一，但他前半生在芬兰的设计实践依然具有无可替代的影响力。他的儿子小萨里宁是20世纪享誉全球的几位天才建筑大师之一，但小萨里宁的设计生涯基本在美国，因此只能被看成芬兰建筑学派的编外人员。芬兰虽然在地理上远离当时欧洲文化艺术的核心地区，如法国、比利时、英国、奥地利等，但以老萨里宁为代表的一大批芬兰建筑师和艺术家依然能将当年欧洲最时尚的工艺美术运动和新艺术运动的创意思潮及时引入芬兰，并随时结合本民族文化传统，创作出一大批芬兰民族浪漫主义风格的建筑经典。即使以今天的眼光和使用标准来衡量，老萨里宁和他那一代芬兰建筑师的经典作品，如赫尔辛基中央火车站、芬兰民族博物馆以及一大批办公楼和公寓，都是建筑史中的典范之作。芬兰建筑师们在悉心秉承欧洲古典建筑传统的同时，也将芬兰和北欧的独特文化因子融入作品当中，基于一脉相承的“抽象与几何”思维模式，老萨里宁将欧洲古典柱式与北欧自然元素转化而成的装饰主题融为一体，创造出宁静、温馨、生态、健康、以人为本的宜居宜用空间系列，开创了北欧人文功能主义学派。

芬兰建筑学派真正名扬世界并开始对全球建筑师产生广泛影响的始于阿尔托，他被著名建筑史学家吉迪翁誉为现代建筑五大师之一，是现代建筑和现代设计的一代宗师。与另外四位开创现代建筑新面貌的建筑巨匠——赖特、格罗皮乌斯、密斯和柯布西耶相比，阿尔托相对年轻，相对而言没有其他几位那么多的开创性举动，但阿尔托却是集成设计的顶级大师，同时他更关注社会、关注人类、关注环境生态和可持续建筑发展，是影响极为深远的建筑思想家。芬兰建筑学派也因此具有与早期现代主义建筑不同的思维特质，并能在整个20世纪都能保持其原创思想的影响力，以及对生态环境始终如一的关注。阿尔托被赖特誉为“一代天才”，因为他能够将古希腊、古罗马的经典柱式和同时代建筑同仁的先锋理念融会贯通后，再以极具创意的手法开创完全属于自己和芬兰的设计语言。他用广收博取的气度兼收并蓄欧式几何和北欧几何的数学语法及语汇，再以立足于芬兰本土的建筑材料为核心，构筑起服务于基本功能但又超越基本功能要求的抽象空间系统。阿尔托的建筑千变万化，但始终令人百看不厌，使用者身处其中，都会从内心生出赞叹。阿尔托的建筑天才还表现在建筑原型的开创性研究上，在疗养院与医院、图书馆、剧场、大学、教堂、住宅与建筑类型的研究和设计方面，阿尔托均有开创之功。此外，阿尔托更是现代家具和工业设计领域最重要的开拓性大师之一，对现代设计的创意观念具有革命性的影响。

众所周知，天才不可复制，或难以复制。现代建筑的天才大师如赖特、柯布西耶和阿尔托等，开启了一代设计新风，但后世却很难再产生这样的一代宗师。因此，优秀的建筑学派若希望能够持续产生优秀的建筑师，必须严肃思考建筑研究和设计教育的问题，在这方面，芬兰建筑学派的幸运就在于在阿尔托之外，还有布隆姆斯达特。

笔者曾经在早期的一篇论文《芬兰建筑的两极》中，系统介绍了阿尔托和布隆姆斯达特这两位对20世纪芬兰建筑影响最大的建筑导师。阿尔托的影响主要体现在以自己的天才成就树立榜样，建立文化自信，然而阿尔托终其一生只有极少岁月的教学时间，他以作品说话，却难以哲学理念系统言说。布隆姆斯达特则在诸多方面与阿尔托相反互补，他虽然也时常开展设计实践，因为这是芬兰建筑教授的传统，但他绝大多数时间在于系统教学和对建筑比例与和谐尺度的研究。布隆姆斯达特学识渊博，深知建筑教育对芬兰建筑持续发展的意义，与此同时，关于提倡什么样的建筑教学，更是他呕心沥血、终生研究的主题。

芬兰当代建筑大师和享誉全球的建筑教育家和评论家帕拉斯玛告诉笔者：在芬兰现代建筑的宝库当中，有两座似乎永远挖不完的“金矿”，一座是阿尔托，另一座是布隆姆斯达特。布隆姆斯达特天赋禀异、思想深邃、见识卓越、影响久远，他的名言：“如果你希望达到最现代的，那么你必须了解最古老的。”笔者早年撰写相关文章时，曾数次前往芬兰国家建筑博物馆中的档案馆，初窥布隆姆斯达特这座“金矿”的奥秘。这位大师一生功业中基础研究的深度和广度，在某种意义上可以与阿尔托的设计事业相提并论。在20世纪的建筑模度研究中，柯布西耶和布隆姆斯达特是取得最高成就的两位设计导师，前者的研究伴以大量成功的设计实践，影响全球建筑界；后者则从音乐入手，纵论古今，开创芬兰建筑一代新风，成为芬兰建筑学派昂然屹立至今的原动力。在芬兰国家建筑博物馆的档案馆中，笔者有幸亲眼所见布隆姆斯达特研究模度系统中比例、尺度与和谐关系方面的绘图原作，看到他从古埃及的建筑比例和人体尺度系统入手，绘出数十幅彩色分析图，看到他仔细整理意大利文艺复兴时期从阿尔伯蒂到帕拉第奥诸位大师的柱式研究手稿，看到他分析并绘制欧洲巴洛克时期著名音乐家的乐谱分析图，笔者心中开始真正理解芬兰现代建筑的力度和温馨之所在。联想起前几年在欧洲各地相关展览中看到的达·芬奇笔记手稿、克利绘画研究手稿和毕加索人体比例研究手稿等，笔者深深体会到深度研究的强大力量，并再次回想起笔者的导师郭湖生教授在谈到学术研究时对我们的教诲："没有广度就没有深度。"

布隆姆斯达特的模度研究博大精深，影响深远，他的设计实践精雕细琢，堪称典范。因此，建立在这种研究和实践基础上的建筑学教学则充分体现出其有效、有力、有温度。可以这么说，在芬兰建筑界，阿尔托的影响是灵感性的，布隆姆斯达特的影响是系统性的。前者难以传承，带有随意性；后者则通过系统教学和研究，带有强烈的传承性。因此我们可以看到，这两位大师身后的芬兰建筑界，能够继承阿尔托灵感衣钵的屈指可数，而绝大多数芬兰建筑师都受到布隆姆斯达特教学系统的熏陶和训练。布隆姆斯达特的建筑教育强有力地确立了芬兰建筑学派功能主义和人本主义发展的主流，但在他生前和生后，依然能够看到芬兰建筑的多元化生态，这也从另一个视角看到其教学的生态之处。

在阿尔托和布隆姆斯达特两峰并立的芬兰建筑黄金时代，也有一批极具创意潜力的芬兰建筑师脱颖而出，其中最突出的当数比尔蒂拉和路苏沃里，这两位建筑大师的共同特征是教学、研究和设计实践并重。比尔蒂拉多少有一些如阿尔托天才型、灵感型建筑师的影子；而路苏沃里则长期作为布隆姆斯达特的助教，坚定而牢固地继承着芬兰建筑人本功能主义的特点。与阿尔托不同的是，比尔蒂拉酷爱写作和研究，尤其喜爱从完全不同的文化背景出发创作不同风格的建筑作品；而路苏沃里则在很大程度上追随布隆姆斯达特的研究与教学方法，但却在设计实践中发展出与导师完全不同的混凝土建筑风格。路苏沃里的建筑作品严谨有力，同时又富于情感和活力，充分体现出芬兰式模度系统研究的力量。实际上，路苏沃里在20世纪60年代创作的那批混凝土建筑，不仅为芬兰建筑开创一代新风，也对包括安藤忠雄在内的一大批国外建筑师影响巨大，而其建筑当中所蕴含的"抽象与几何"思维模式的力量，是这种广泛影响力的根源所在。

比尔蒂拉和路苏沃里之后的芬兰建筑学派开始呈现多元化趋势。其中有以莱维斯加为代表的以音乐模度导入设计理念的芬兰本土风格，有以海林为代表的呼应国际高技派的芬兰高技派风格，有以萨米宁为代表的材料研究引导设计创意的芬兰个性化风格，有以皮罗宁为代表的芬兰钢结构风格，有以帕拉斯玛为代表的以学术研究为导向的芬兰学者式设计风格等。然而，所有风格都在很大程度上坚定捍卫着"抽象与几何"的创意思维理念。事实上，对"抽象与几何"的极致坚持是芬兰绝大多数建筑师的立身之本，而海基宁与科莫宁是这方面的典范。

5．海基宁与科莫宁：从"抽象与几何"到建筑杰作

生于20世纪40年代的米高・海基宁和马可・科莫宁，20世纪70年代中叶毕业于赫尔辛基理工大学（现阿尔托大学），其成长的年代正值阿尔托主宰芬兰建筑的最后阶段。欧洲现代建筑的发展，由德国和法国发轫，以严酷清冷的国际式风格著称，直到阿尔托接过现代主义的接力棒，通过地域主义和以人为本的理念，建立现代建筑富于人性化的风格，并由此将芬兰推向现代建筑运动的引领地位。然而在芬兰国内，随着经济发展所带来的数量巨大的住房问题，以及高科技在建筑中的加速需求，阿尔托所表达出来的个性化因素开始受到质疑。在这种质疑中，布隆姆斯达特和路苏沃里开始建立一种完全基于"抽象与几何"思维模式的新理性主义建筑观念。与此同时，比尔蒂拉从阿尔托的地域主义理念出发，以其丰富的理论和实践建立起一种建筑生态学设计理念。正当阿尔托享誉全球时，他在自己祖国的主导地位开始受到多方面的挑战，这正是芬兰建筑能够不断进步并始终在全球保持高水准的重要原因：一种富有活力的文化不会在长时间只受一种思潮的影响；一种文化，如果没有矛盾，没有挑战，那它就会很快面临衰亡。

当海基宁和科莫宁开始工作时，芬兰的总体建筑风貌正处于一种多元化时期，被国内外评论界冠以"地域主义""人性主义""浪漫主义""浪漫现代主义""新理性主义""生态主义"等特征，海基宁和科莫宁正是在这种背景下开始建立自己的设计理念，并最终走向"人文功能主义"。

海基宁和科莫宁的成名作是1986年通过芬兰全国建筑竞赛赢得项目的“芬兰科学中心”。该中心1989年正式对外开放，这让建筑师也一夜成名，他们在该项目的设计过程中树立起严谨的科学探索精神和对设计细节精益求精的工作态度。该项目在当时极具挑战性，不仅因为方案本身所包含的全新理念，而且在设计的各个阶段都需要反复研习数学、光学、声学、电学、生物学、化学、天文学、地理学、医药学和博物学诸多领域的内容。由此使他们更加坚定地认为：建筑是一种对大自然的矛盾与协调进行统一考虑的设计整体，建筑师要从自然万象的纷呈混沌中寻找有规律且平衡的力量并加以提炼，如同科学研究一样。海基宁非常欣赏法国哲学家保罗·瓦拉里的一句格言：“有两样东西从未停止过干涉这个世界，即无序和有序。”在大自然中无处不在的就是这种人类眼中的无序和有序、预感和随机，它们总在冲突与调和中。对于建筑设计而言，这种源自科学研究的态度实际上蕴含着无穷的审美潜力。无序需用抽象来审视，有序则依赖于几何模式去解读，由此形成一种设计的体系。大自然本身是一种平衡而美妙的整体，在人类有限的思考能力范围内，大自然是奇异的，又是和谐的，一瞬闪电雷鸣，一道雨后彩虹，每日的朝霞万道或落日余晖，都会对人类的审美思维产生不可抗拒的影响。海基宁和科莫宁从其成名作“芬兰科学中心”开始，立刻陷入对“抽象与几何”思维模式的深度迷恋，并自然延续到其后的每一个建筑项目中。

对海基宁和科莫宁而言，建立在“抽象与几何”基础上的科学思维不仅不会枯寂，还可以变幻无穷，充满情趣，因此，他们的建筑总是兼具严肃性和幽默感，时常能在节制与随意之间畅游。对抽象和几何在任何一个层面上的充分理解，实际上都可以在建筑空间的创造上产生无穷尽的变幻模式。“芬兰科学中心”的功能序列即由立体几何学中的不同元素组合而成，其间再用不同材料的不同节点切入贯通，纯几何学中的立方体和球体在此以各种变体形式演化出不同的功能空间，其本身就是对科学的一种空间形态意义上的解读。

海基宁和科莫宁有一个信念，即建筑并非风格的追逐，而是关于诗意的艺术。他们在学生时代和后来的设计实践中，日益坚信对抽象与几何思维的执着至极致就是一种诗意，是建筑设计的主旋律，它无处不在地贯穿于海基宁和科莫宁建筑作品的始终。他们的建筑中几何体量的意向构成往往精炼至极，如同康定斯基、马列维奇和莫霍利-纳吉的绘画与装置艺术，突显诗意。芬兰罗瓦涅米机场候机楼的立方体组合，意为用几何特征强烈的金属盒子表达一种寒冷地区下的永恒气息，同时也展示极简主义的工业美学；芬兰库比奥急救学院则将培训人们平静面对火灾和危难境地的理念反映在建筑总体布局中，在冷酷的几何形体组合中，狭长的教室部分与月牙形宿舍楼形成鲜明对比；而在位于丹麦的欧洲电影学院，海基宁和科莫宁更是用纯几何形体的组合完美地解决了所有的功能问题；在华盛顿的芬兰驻美国大使馆，两位成熟的芬兰建筑师用最时尚的金属和玻璃诠释几何形态下的诗意，其外观立方体的庄重与室内金属楼梯的奔放仿佛保守沉默的芬兰人有礼貌地掩饰着自己火热的情感，该馆1994年开馆典礼时，建筑大师贝聿铭曾专程前往祝贺。严谨而彻底的几何形体，以及与不同基地环境的协调处理，构成海基宁和科莫宁作品的重要基石，就像赫尔辛基沃州里文化中心巨大的混凝土半圆弧，其中非常温馨地容纳着一个图书馆的文化活动中心。在其近几年完成的建筑项目中，如位于德国德累斯顿的系统生物学中心，位于瑞士卢塞恩的斯托布尔公园，芬兰萨文林那图书馆等，海基宁和科莫宁在设计中融入更多高科技成果并积极参与新型材料和建筑配件的研发。这种研发时常体现在其非常频繁而精准的模型制作中，他们的每一个设计项目对他们而言也是一种科研探讨。海基宁和科莫宁的建筑，表面上是在以各种不同组合重复使用着直线与圆弧、立方体与球体，配以各类金属屏幕、各种玻璃和其他工业产品，以此做一种形式主义的抽象游戏；但实质上，他们始终追求的却是在对纯形式理想的探索中达成一种诗意，并在这种语言中完成充满理想主义色彩的人文功能主义设计理念。

西方建筑的发展，从古埃及、古西亚到古希腊、古罗马，历经中世纪再到文艺复兴，而后再经巴洛克、洛可可、新古典主义、工艺美术运动、新艺术运动和新装饰运动而最终进入现代主义，其不变的主题是对更大、更多、更人性化空间的追求，其永恒的理想则是用抽象的思维和几何的尺度展现一种诗意。从赖特和老萨里宁，到柯布西耶、格罗皮乌斯、密斯、阿尔托和布劳耶尔，再到路易斯·康、贝聿铭、布隆姆斯达特、比尔蒂拉和路苏沃里，这批现代主义建筑的桂冠诗人们以其惊世才华谱写着一曲又一曲现代建筑的优美篇章，在这众彩纷呈的篇章中，“抽象与几何”是其创意设计中不变的主旋律。如何在新时代激发创造的诗意？如何在当今全球信息时代保持并弘扬现代建筑运动中人文功能主义的内核？同时作为建筑师和建筑学院教授的海基宁和科莫宁始终都在思考着这些问题，并试图用自己不懈的设计实践做出某种意义上的回答。立足于“抽象与几何”的创意模式，他们永远会以非常执着的精神探索现代建筑的潜力，即人类的建筑究竟能在多大程度上阐释人与大自然的关系。他们的作品一方面是对纯几何形体近乎狂热的追求，另一方面又对所有新材料与新技术全盘接纳，两者的结合便自然形成了一种弥漫于建筑中的对人类、对环境、对未来前途的乐观态度。

人类社会在过去五十年的发展可谓日新月异，当今时代科技水平的飞跃从宏观到微观诸层面都速度惊人，但建筑上所谓的高科技充其量大都依然停留在蒸汽机时代的水平上。建筑师不应该轻易满足，而应以更执着的心态追求和完善“抽象与几何”的创意模式，并同时密切关注科技领域的进步和生态学的发展，唯有如此，方可达成诗意的创造和美学的升华。海基宁和科莫宁的作品展示出一种坚实的美感，这种美感来自抽象的简约模式和对几何形式语言的精确专注；他们的作品充满极具创意的思考，这种思考包括对各种文化思潮的有意识和无意识的吸收，更包括对自然环境、科学理论、建筑结构和材料细节明智的欣赏和体验。

模型草案

米高·海基宁

在过去，芬兰建筑师的职业生涯通常是从建筑设计事务所中制作建筑模型的初级助理岗位开始的。在我的第一份工作中，我很快就发现自己要学会用斯坦利刀切割纸板。当然，我曾经为我的铁路模型制作过飞机模型及石膏景观模型，但是我对建筑模型知之甚少。制作模型的材料阻力非常大，对软木棍进行纵切很不容易，并且胶水会从接缝处流出。当时正在做的模型立面应该是一种由细金属棒制成的模块化框架结构，在当时很流行。我从位于我们办公室附近院子里的一个小车间订购了这些金属棒，而这些金属棒可能是我的模型结构中唯一精确的配件。然后，我有幸把我的“作品”带到了项目决策会议上，会议地址就是由阿尔瓦·阿尔托设计的位于赫尔辛基的萨沃伊（Savoy）餐厅。我当然听说过这位传奇建筑师的这座杰作，也急切地等待着能够在现实中见到它。在门口，一位冷漠的门卫瞥了一眼身穿网球鞋和破旧美国军队外套的我，道了声“谢谢”，就从我手中拿走了模型，并关上了门。

从那时起，建筑事务所的日常生活彻底发生了改变。数字化工作都是在计算机屏幕上完成的，而且更加逼真的3D建模软件甚至能挑战摄影。然而，硬件和软件的发展似乎并没有取代传统建筑模型的作用。新材料和新设备，如聚苯乙烯泡沫塑料、激光切割和3D打印，已经创造了诸多新的可能性。

许多知名建筑师都承认，模型已经成为其设计过程中不可缺少的工具，尤其是在项目的最初阶段。建筑师彼得·卒姆托用模型照片的图像展示他的想法，他的工作室堆满了各种大型的建筑模型。这些模型在他与助手的讨论中和向客户所做的展示中发挥着重要作用。这些模型都是由富有创造性的、令人惊奇的板材制作出来的，看起来很像艺术品。

日本建筑师藤本壮介[1]（Sou Fujimoto）和他的设计团队通过模型来相互交流与沟通，其制作的模型数量巨大，2014年伦敦的蛇形画廊展馆中就展出了近70个模型。有时在完成100个模型版本之后，他们可能又会将前面的方案推倒，从零开始。例如，他曾表示，“在设计的早期阶段，概念性草案模型能够将我们的想法及讨论的话题，或者可能的建筑体量……可视化”。

葡萄牙艾利斯·马特乌斯（Aires Mateus）兄弟建筑事务所认为：“我们对模型的兴趣是一种现象学意义上的痴迷。”他们在项目开始时就会建造大型模型。他们发现，微型模型的弊端在于即使一个丑陋的解决方案也可能看起来很不错。“我们喜欢在设计作品的时候就好像是从内部观察它，这就是为什么我们的模型通常按照1∶20的比例来制作。”

在日本东京的SANAA建筑事务所[2]，其中一个项目是以模型为中心而展开的。设计团队中的每个人都使用模型，通常没有任何图纸，讨论也围绕模型进行。西泽立卫认为：“看着模型，你可能会认为这个想法不好，或者这个想法比我想象的更好，它为我们的下一步设计打开了思路。”

在瑞士建筑大师雅克·赫尔佐格和皮埃尔·德梅隆[3]举办的名为“自然历史”展览的目录册中，概念模型或者说材料研究，展示了物质上的变化是如何彻底改变形式特征的。模型的透明度、制造工艺、形式精度和纹理，都可以引导设计过程。

20世纪90年代末，我前往位于美国纽约市哈德逊街上斯蒂文·霍尔的办公室。接待处的墙面上挂满了斯蒂文·霍尔自1992年以来为芬兰赫尔辛基当代艺术博物馆竞赛所制作的手掌大小的模型。乍一看，它们似乎没

1 藤本壮介（1971— ）是日本新生代最有才华的建筑师之一，曾获建筑评论奖，2008年获日本建筑家协会奖，2008年获巴塞罗那世界建筑节一等奖，2009年获Wallpaper奖。藤本壮介经常用“原始”形容他的作品，他把建筑实践看成是探索世界和人道的一种方式。——译者注

2 SANAA建筑事务所由妹岛和世和西泽立卫于1995年共同设立。2004年获威尼斯建筑双年展金狮奖。2010年获普利兹克奖。——译者注

3 瑞士著名建筑团队。1978年，德梅隆先生和赫尔佐格先生合伙建立了一家建筑设计事务所，1997年正式采用赫尔佐格和德梅隆的名称。2001年获普利兹克奖，其在中国的设计为2008年奥运会中国国家体育场方案——“鸟巢”。——译者注

有任何共同之处，但再次观察后，你会发现一个发展脉络，这最终使该设计在比赛中脱颖而出。

经过“评审和否定”过程后，许多模型将面临突如其来的“死亡”，并最终被扔进工作台下的垃圾箱，而有些幸存的模型在设计过程的演变中作为曾经的存留仍然存在着价值。对于未知的将来项目，这些模型可能会包含新的解决方案或新方法，并且可以在完全不同的环境下被重新使用。在比亚克·因格尔斯（Bjarke Ingels）的BIG建筑事务所[1]的办公室里，存满了他设计的所有旧模型，它们作为一个创意银行有可能有新的用途。

[1] 比亚克·因格尔斯是BIG建筑事务所的创立者、设计总监。BIG是一个由建筑设计师、产品设计师、设计思考者组成的，位于丹麦首都哥本哈根的国际化事务所，在建筑、城市设计等多领域均有广泛的合作实践。——译者注

无论是储藏室的模型还是玻璃橱窗中的模型，都展示了创造者为寻找最佳解决方案而付出的不断努力。在“自然历史”展览的目录册中，雅克·赫尔佐格和皮埃尔·德梅隆提出了这样一个问题：模型是建筑本身，还是只是设计过程遗留下来的废料？并给出了他们的回答：“这些被保存的模型就是毫无价值的废料，因为在这类非物质建造过程中，理解、学习和发展总是要处于优先地位。”

长期以来，我们办公室里的建筑模型大多是为展示最终设计作品而制作的，也有的是为参加建筑设计竞赛，或是为满足客户需要制作的。草图设计主要是用笔和纸来做，自从设计过程由开放式工作台转移到个人电脑屏幕之后，设计项目在办公室的物质存在就消失了。通过使用图像渲染程序，可以创建3D模型，并可以在电脑屏幕上进行旋转，但图像仍然是二维的，而且每个人看到的都一样。而使用草案模型似乎弥补了缺失的一环，即团队所关注的焦点。通过一个真实的等比例模型，项目的进展将以一个具体的实物形式呈现，整个设计团队可以参考它，无论是用于获取灵感还是超越挑战。SANAA的西泽立卫曾说：“如果你有一个想法，你就做一个模型。它会把你的想法变成一个实物，并且可以让人看到。”

就我个人而言，用模型进行草案设计和借助模型进行思考，已经取代了手绘草案的过程。在一个项目的初期阶段，使用铅笔画图已经简化成为

模型设计的准备阶段。建立模型需要某种想象力，这也是解决设计任务的一种理念。一个草案模型会迫使你将自己的想法转化为一个可视并且可展现的形式。显然，用模型来实现并不清晰的想法肯定是困难的。有时候，相比语言，模型可以更好地展示我的思路，以及那些我打算知道但还未知的东西。

在2010年中国上海世博会芬兰馆的建筑设计竞赛中，我们的想法起源于办公室附近公园的一棵树桩。我在森林里搜寻了一棵大小合适的被砍伐下的白桦树，从上面锯下一块薄薄的圆木，并将其雕刻成所需的形状，再用镜面膜覆盖住切割面。一眼看去，这块木桩展现出森林般的外观，而其内部则像一个现代化的展览馆。

在芬兰驻日本东京大使馆的建筑设计竞赛中，我们想把建在同一地块上的大使馆大楼和住宅公寓“融合”成一体。通过折叠薄薄的描图纸，很容易创建出雕塑般、几乎无缝的实体模型，其整体形式可以通过下面的灯光得以突出。基于由缩小尺寸的草案模型和鸟瞰透视图共同呈现出的像实物一样的效果，方案可以得到进一步的完善。而且根据3D模型拍摄的二维照片能够呈现建筑物全方位的视觉效果。

手工艺品商店货架上的产品激发了我的思考：“我想知道用它们可以做些什么。”最有趣的材料在任何地方都能偶然发现。我们的一个模型制造商表示，他从垃圾场获取灵感，并从那里获取物资，以备将来使用。通过选择不同的材料，即使相同的形式也可以呈现出完全不同的特质。3D打印机是展现模型体量和复杂构件的极好工具，但它无法展现材料的属性（例如不同程度的透明度和反射性）。

通过使用拼贴技术，很容易在一个模型中将白色纸板转换成如陶瓷图案的混凝土。在位于芬兰赫尔辛基的弗洛拉诺基奥（Flooranaukio）住宅项目中，建筑物外墙使用来自邻近的阿拉比亚（Arabia）陶瓷厂废瓷片的想法，引发了诸多实验。例如，把从互联网扫描到的陶瓷纹理粘贴到1∶200比例模型的表面上，与20世纪30年代这个工厂生产的花瓶装饰图案效果是一致的。

有时候，模型可以是在设计方案过程中展示基本思路的快捷方式。在芬兰赫尔辛基阿尔玛（Alma）媒体总部的设计过程中，我们只用一个晚上就想出了针对特定设计问题的解决方案，包括大楼的分隔、沿着场地的边缘和周围建筑物的方向排列。亚克力板代表不同的楼层，而楼层间采用木销穿孔方式连接形成楼梯部分。建筑物的形状——立面上实体表面与开口形成的条状关系，通过投影仪幻灯片得以实现，幻灯片上印有媒体大楼的各种品牌名称。比例模型的照片，可以创造出建筑物真实特征的视觉效果。

对于位于芬兰耶尔文佩（Järvenpää）的珀赫拉（Perhelä）住宅区的竞赛设计方案，在比例模型的帮助下，我们研究了由垂直部件组成的建筑物的视觉效果是如何依据观看视角和照明而改变的。对比例模型拍摄的一系列照片，从类似于微型雕塑的模型上展现出一个真实建筑的印象，并且这些照片在竞赛展板中呈现为360度的全景图。

然而，模型有时候在自己的尺度中效果最好——将它在手掌中转动，就像一个类似于要触摸的护身符一样。德雷斯顿（Dresden）系统生物学中心的概念模型，是由坚硬的桃花心木制成的，它代表了该建筑迷宫式的混凝土中心部位。模型类似于日本的库米基拼图（Kumiki puzzle），由于拳头大小的物体上有凹痕，因此，在手掌中不仅能给人一种视觉感受，还能给人一种触觉感受。

本书介绍了我们工作室在参加建筑竞赛或委托项目中的早期阶段制作的一系列模型，大多数模型都是在电脑屏幕或绘图板旁建造完成的，那些基本工具和材料都唾手可得。然而，最终的效果模型均是从外面专业的创作室订购的。建筑模型支撑设计团队余下的一些工作，反之亦然，设计过程也为模型开发提供反馈意见。与任何项目的设计草图一样，大多数概念模型都像废纸一样随着项目的进展而消失，但这些模型的照片仍然保留了设计过程中不同阶段的记忆痕迹。

用模型来做概念

CONCEPT BY A MODEL

系统生物学中心

德国，德累斯顿，2017年

Centre for Systems Biology

Dresden, Germany, 2017

该建筑的结构可以比作对一个问题的解决方案，人们仿佛通过一个复杂而又昏暗的中心走向光明。迷宫式的混凝土中心结构为研究人员的相遇和互动提供了舞台，并为科学实验提供了便利。

为了研究该建筑外围与迷宫式中心之间的比例关系，我们按照 1 ： 500 的比例制作了木质概念模型。庞大的中心部分可以从模型中移出，让空间概念更加视觉化。不同种类的木材用来区分外围的办公区域与建筑的实体中心。

The architecture of the building is a metaphor for the solution to a problem. One passes via a complex and dim core towards light and clarity. The concrete labyrinth sets the stage for the encounters and interactions between researchers, and provides facilities for scientific experiments.

A wooden conceptual model in the scale 1 : 500 was made in order to study the proportions between the outer perimeter and the labyrinthine core of the building. The massive core unit was removable from the model so as to visualize the spatial concept. Different wood species were used to differentiate the office zone on the outer perimeter from the solid core of the building.

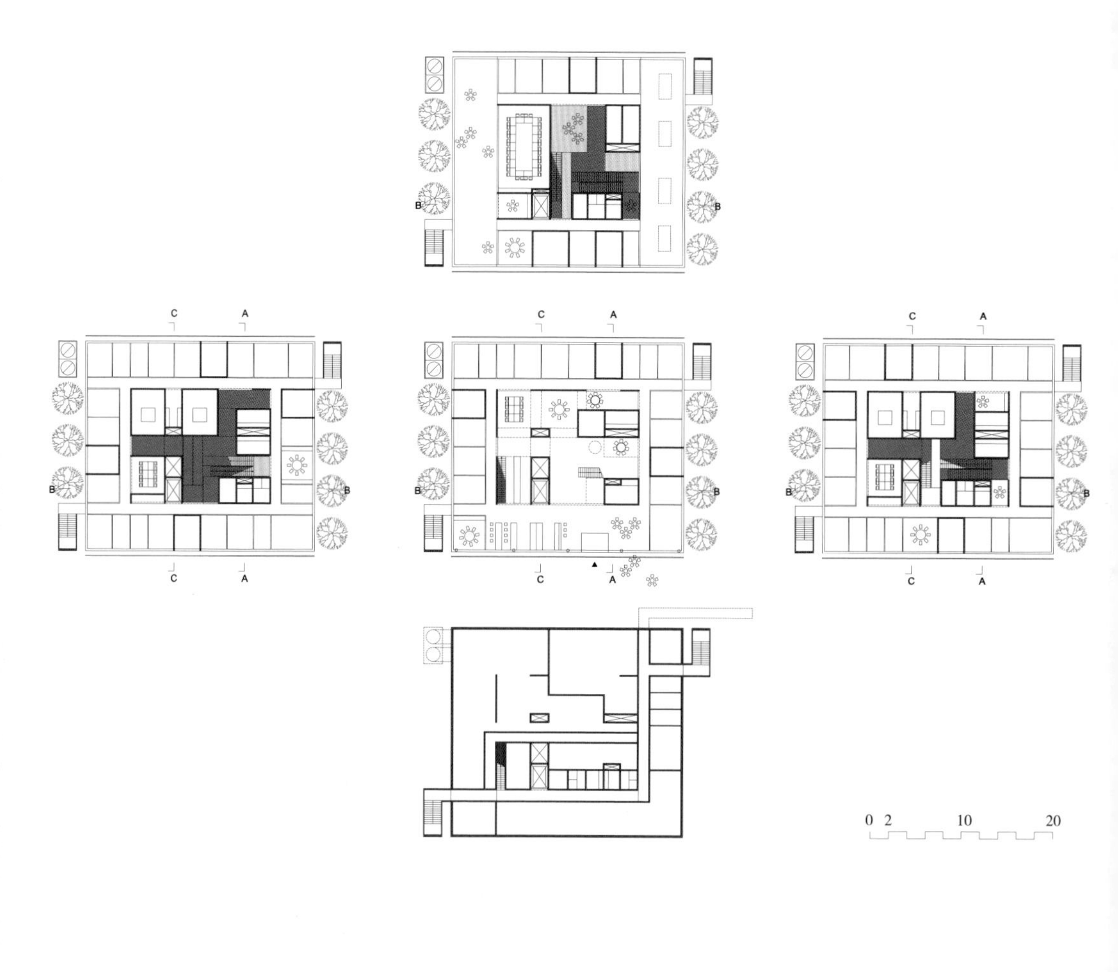
B
B
C
A
C
A
B
B
C
A
B
B
C
A
C
A
B
B
C
A
0 2 10 20

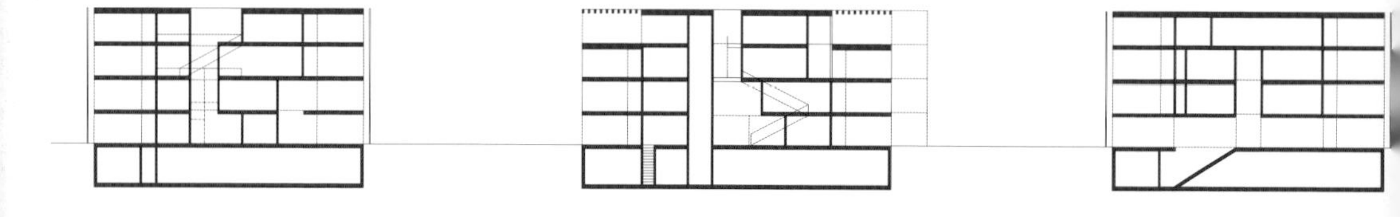

为了更详细地研究空间结构，我们按照 1 ： 200 的比例制作了一个更大的模型。手掌大小的模型可以使建筑目标具体化，同时还是非常有趣的微型雕塑。木制品的密度给人一种结实的、现浇材料的感受。

Also, a larger model in the scale 1 : 200 was made in order to study the spatial structure in more detail. Models that can fit in the palm of your hand can encapsulate the architectural objectives, and at the same time are in themselves interesting miniature sculptures. Due to its density, wood gives the impression of a solid, cast in-situ material.

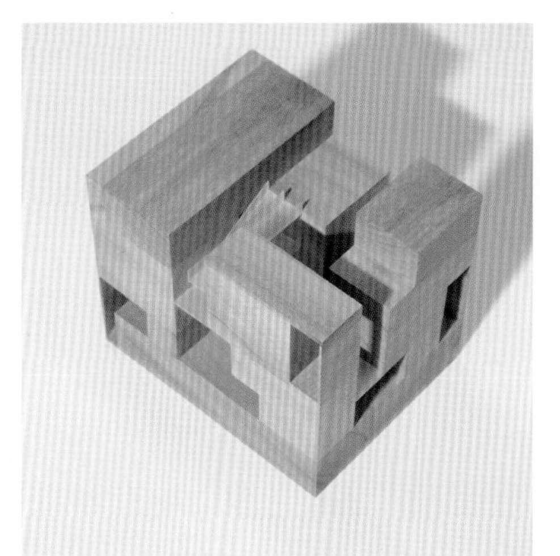

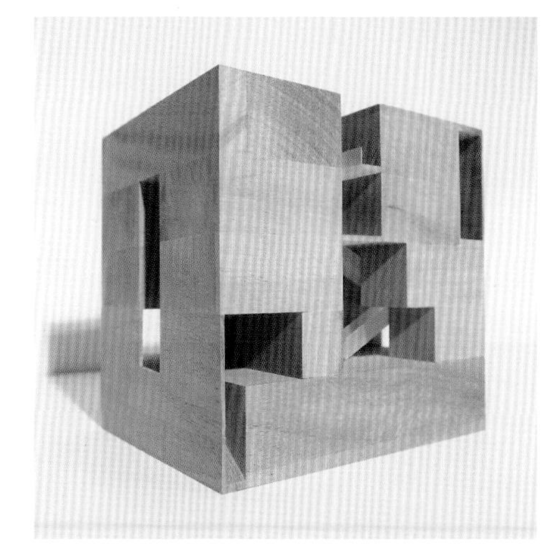
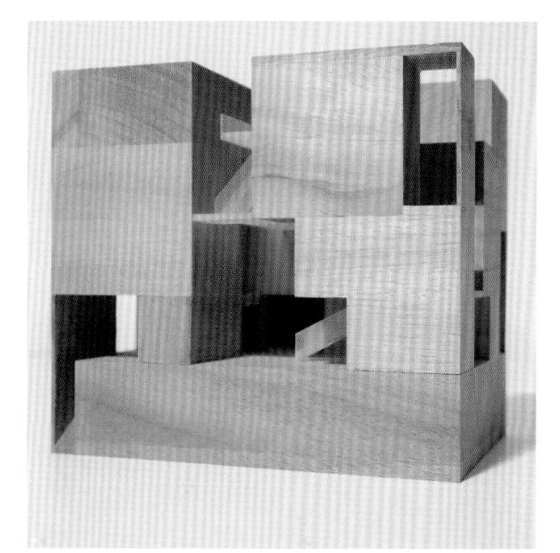

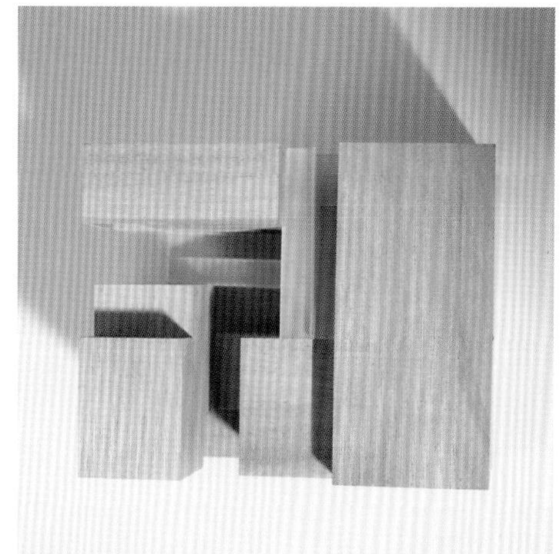

不透明与反射

PACITY AND REFLECTION

芬兰科学中心——赫尤里卡扩建项目

兰，万塔，2016年

innish Science Centre Heureka Extension

antaa, Finland, 2016

建项目的目的是为 1989 年对外开放的芬兰科中心提供更大的展览空间和后台空间。新展厅工作坊的立面由高光泽的铝面板制成，这些板能够将建筑自身的结构以及周边的景观反射来。

了使建筑方案更加视觉化，模型由粘贴了镜面纸的硬纸板制成。之后，在由现场全景照片环的圆盒中，为模型拍照。

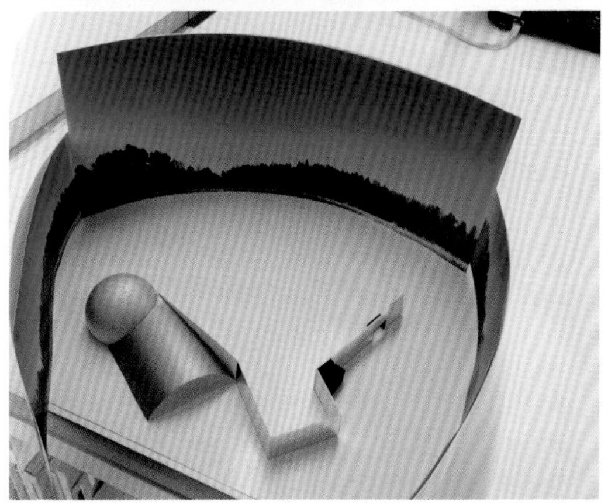

he purpose of the extension was to enlarge the xhibition and back stage space of the Science Cenre, which first opened in 1989. The facades of the ew exhibition hall and workshops were realized ı high gloss aluminum panels that reflect the arhitecture of the building itself as well as the surounding landscape.

n order to visualize the proposal, a model was ıade in cardboard which was then clad in mirored film. It was then photographed in a round ox covered inside with a landscape panorama takn from the site.

在早期的研究中，扩建的部分由一个独立
的自由形体组成。

In an earlier study the extension was com
prised of an autonomous free form volume

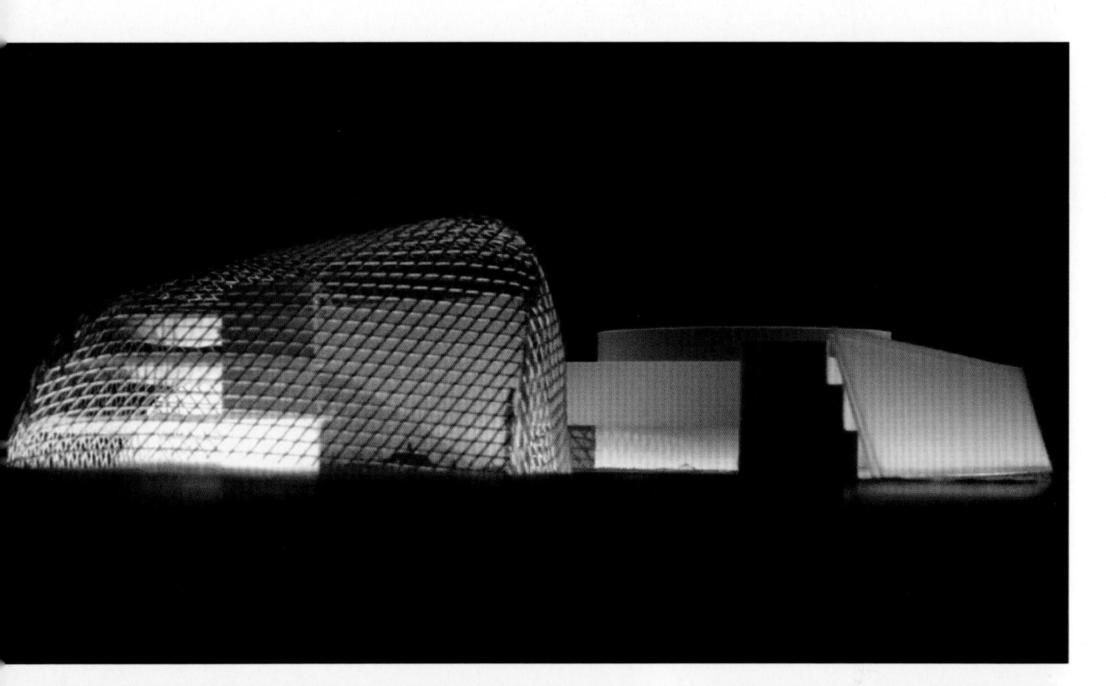

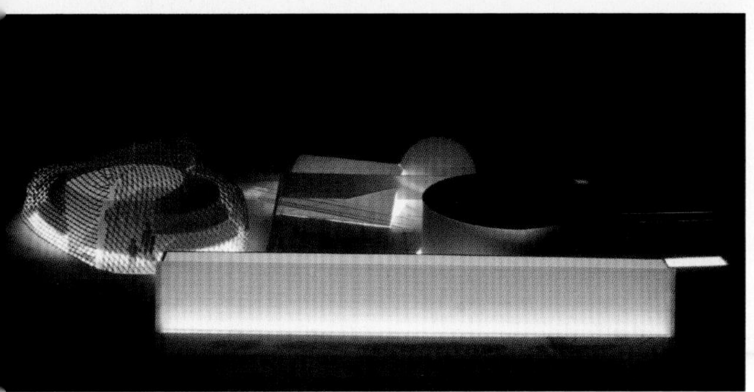

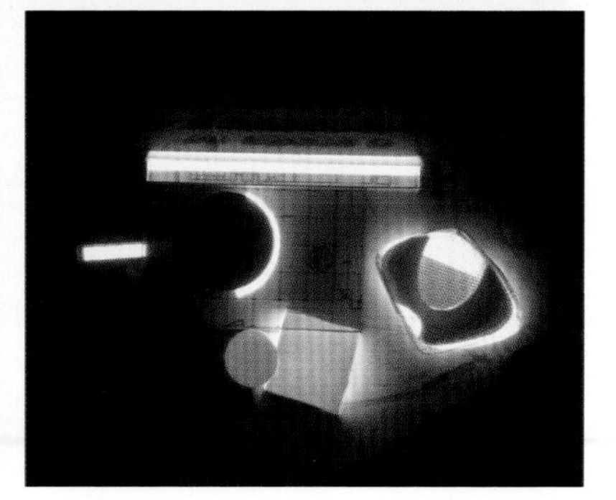

模型由纸板制成，放置在场地和楼层平面图的复制品上。纸板很容易操作，且材料的厚度在接缝处不是特别明显。玻璃的表面都是用条状亚克力板制作的，如入口处的雨篷和面向附近铁路轨道的墙体。而“天文馆”则是从乒乓球上切下来的，尺寸几乎与其完美匹配。扩建部分的自由形状体立面由鱼鳞状形态组成，用金属网呈现，很容易就能弯曲成所需的形状。灯光从模型基板上的切口处射出，从下方照亮模型，而光线的颜色是由放置在光源前的彩色薄片呈现出的。

The model was built from cardboard placed on top of a copy of the site plan and floor plan. Cardboard is easy to work with and the thickness of the material is not particularly visible at the seams. The glass surfaces, such as the entrance canopy and the wall facing the nearby railway tracks, were built from acrylic strips, while the planetarium was cut from a ping-pong ball, the size of which was almost a perfect fit. The free form facades of the extension, consisting of fish-scale-like elements, are shown in metal mesh, which is easy to bend to the desired form. The model was lit from below, through openings cut in the base board. The colour of the light was achieved through the use of coloured films placed in front of the light source.

用模型来渲染

RENDERING WITH A MODEL

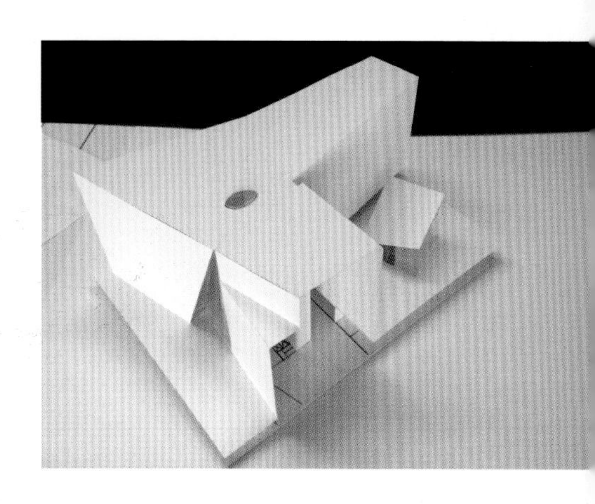

HK4 办公中心

芬兰，赫尔辛基，2015年竞赛，一等奖

HK4 Office Centre

Helsinki, Finland, 2015 Competition Entry, 1st Prize

这座新办公大楼坐落于一个旧工业街区的角落，由多个相互倚靠在一起的立方体组成，其下方是贯穿该街区的人行道和自行车道。外墙的红砖使该建筑与周围环境相得益彰，同时，光滑的白砖表面将日光反射到狭窄的庭院中。

The new office building, located at the corner of an old industrial block, consists of enormous cubelike pieces leaning against each other, beneath which is a pedestrian and bicycle route running through the block. The redbrick of the facades creates a link between the building and its surroundings, whereas the glossy white brick surfaces draw daylight into the constricted courtyard.

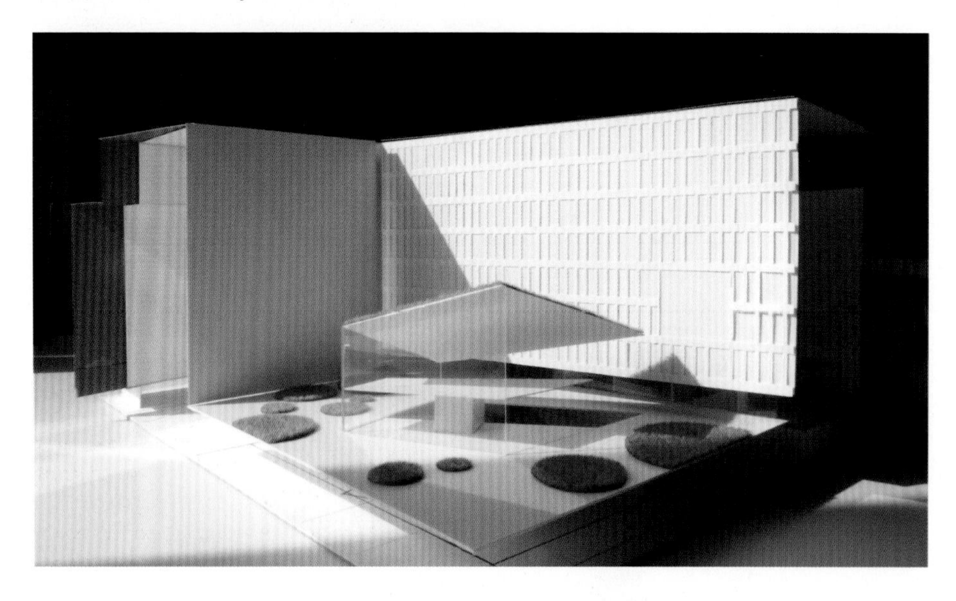

该建筑模型的比例为 1 ∶ 500，由白色硬纸板制成，而比例为 1 ∶ 200 的模型则有更多细节供我们研究穿过建筑物的步行街和主入口。通道表面覆盖着 0.15mm 厚且易于成型的铝薄片。通过在纸板的最外层切割出窗户的轮廓，窗户以浮雕的方式显现。该模型的照片经 Photoshop 软件编辑处理。

The building's massing was explored with a white cardboard model in the scale 1 : 500. With a more detailed model in the scale 1 : 200 we studied both the pedestrian street that cuts through the building and the main entrance. The surfaces of the passage were clad with thin, 0.15 mm aluminium sheet, which is easy to shape. The windows are shown simply in relief, by cutting the outlines of the windows into the outermost layer of the cardboard. The photo of the model was edited in Photoshop.

空洞与体积
VOIDS AND VOLUMES

斯托布尔公园
瑞士，卢塞恩，2014年

Schönbühl Park
Lucerne, Switzerland, 2014

每栋住宅楼都由两个翼楼组成，翼楼之间是玻璃楼梯间。该模型由白色纸板和截面为1.5mm×1.5mm的木条制成。为了确保翼楼部分的坚固性，在公寓窗户的后面粘贴了一张半透明的纸。该模型还展示了典型住宅楼层的房间划分。

Each of the residential buildings consists of tw wings, between which is a glazed stairwell. Th model was built from white cardboard and 1.5×1. mm wooden strips. In order to emphasise the sol id nature of the wing sections, a semi-transparen paper was glued behind the window openings o the apartments. The model also shows the roon division of a typical residential floor.

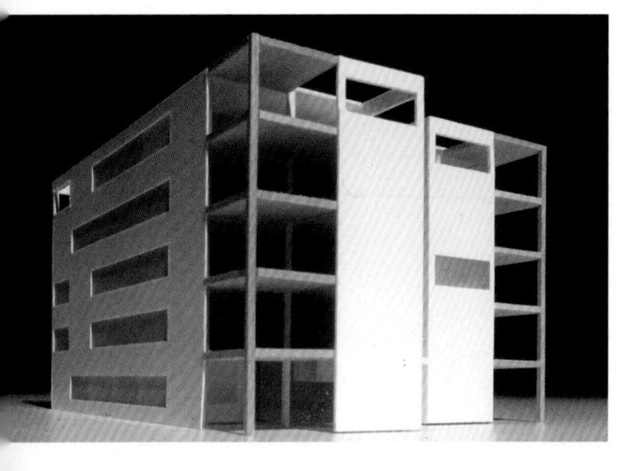

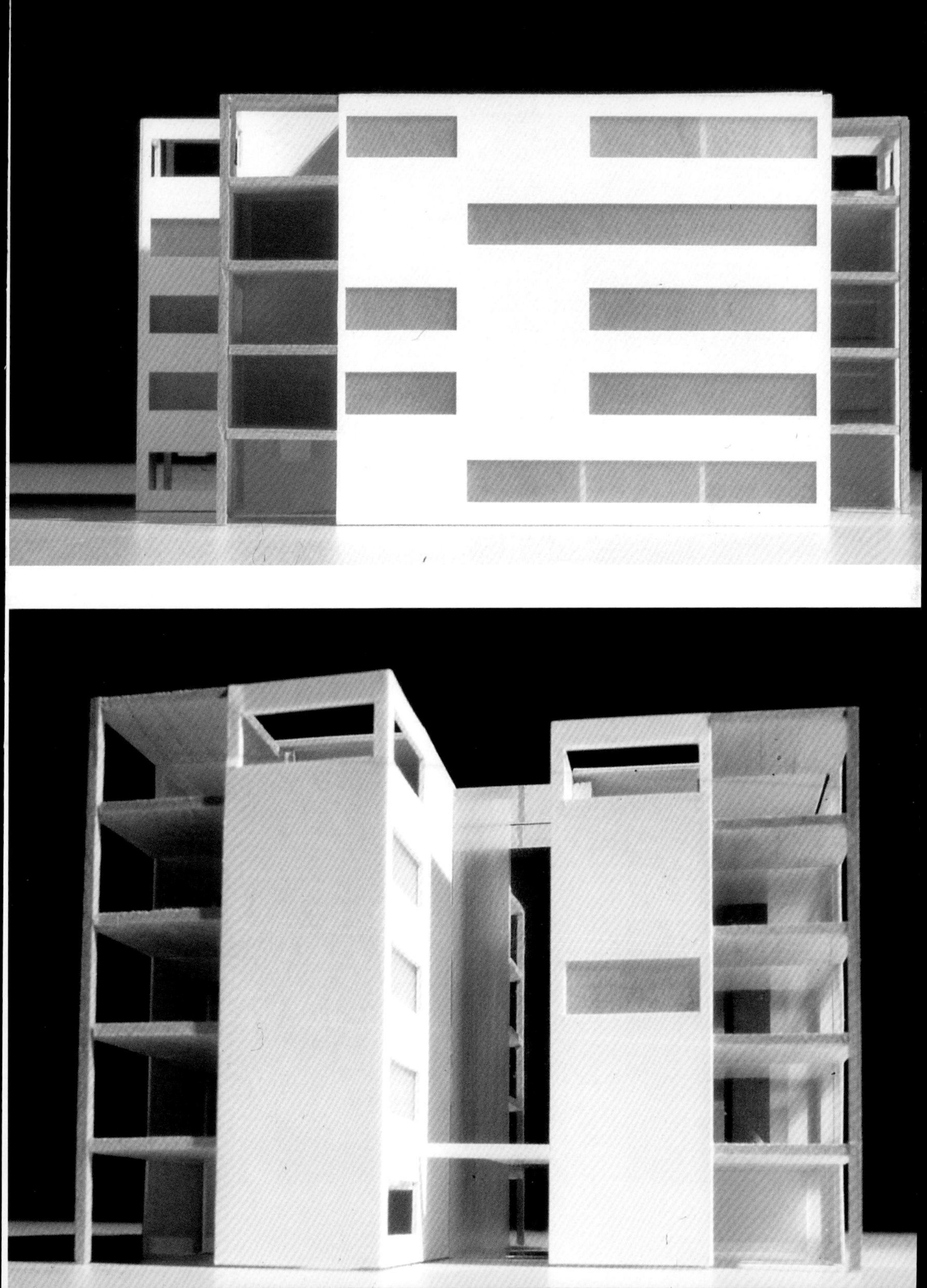

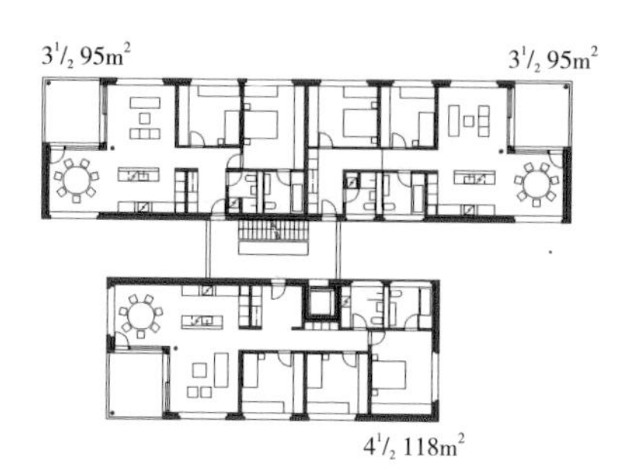
$3^1/_2$ $95m^2$
$3^1/_2$ $95m^2$
$4^1/_2$ $118m^2$

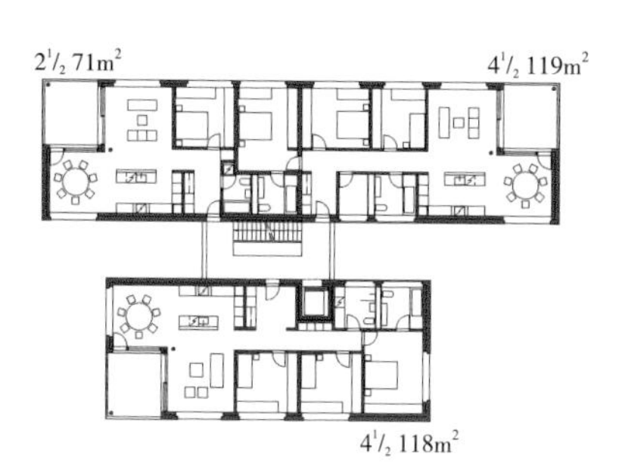
$2^1/_2$ $71m^2$
$4^1/_2$ $119m^2$
$4^1/_2$ $118m^2$

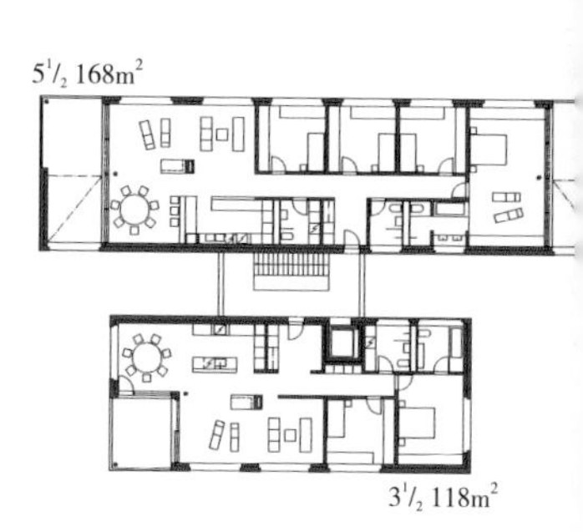
$5^1/_2$ $168m^2$
$3^1/_2$ $118m^2$

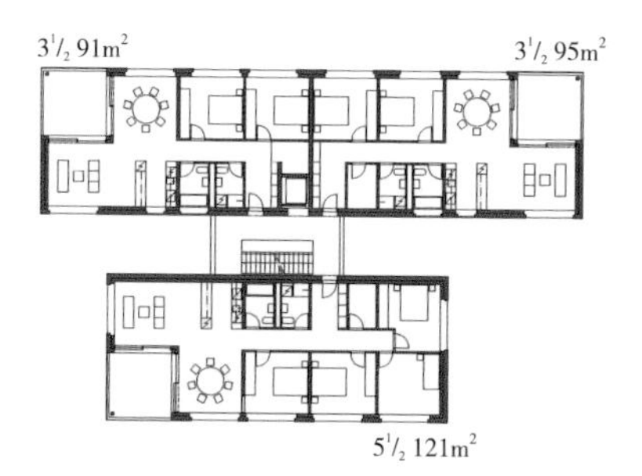
$3^1/_2$ $91m^2$
$3^1/_2$ $95m^2$
$5^1/_2$ $121m^2$

0
5
20

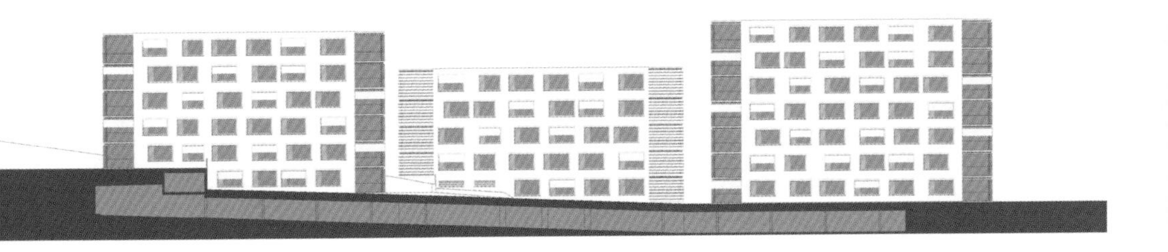

形式追随材料
MATERIAL FOLLOWS FORM

萨翁林纳主图书馆
芬兰，萨翁林纳，2013年

Savonlinna Main Library
Savonlinna, Finland, 2013

图书馆的初稿设计包括勾勒出湖边建筑的尺寸和外观，以及通往城市的一条繁忙路线。草案模型直接在平面图上制作完成，薄金属网格代表了外墙前的一个半透明结构的想法，赋予了该建筑独特的外观效果。

The first draft designs for the library involved outlining the size and appearance of the building next to the lake and a busy arrival route into the city. The sketch model has been made directly on top of the ground plan. The thin metal mesh represents the idea of a semi-transparent structure in front of the exterior wall, giving the building its distinctive look.

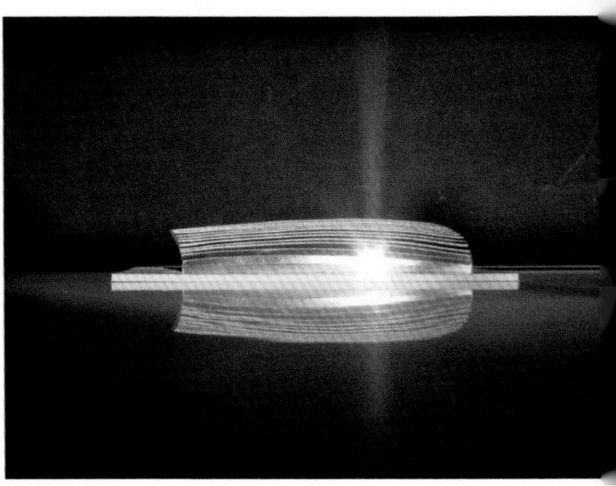

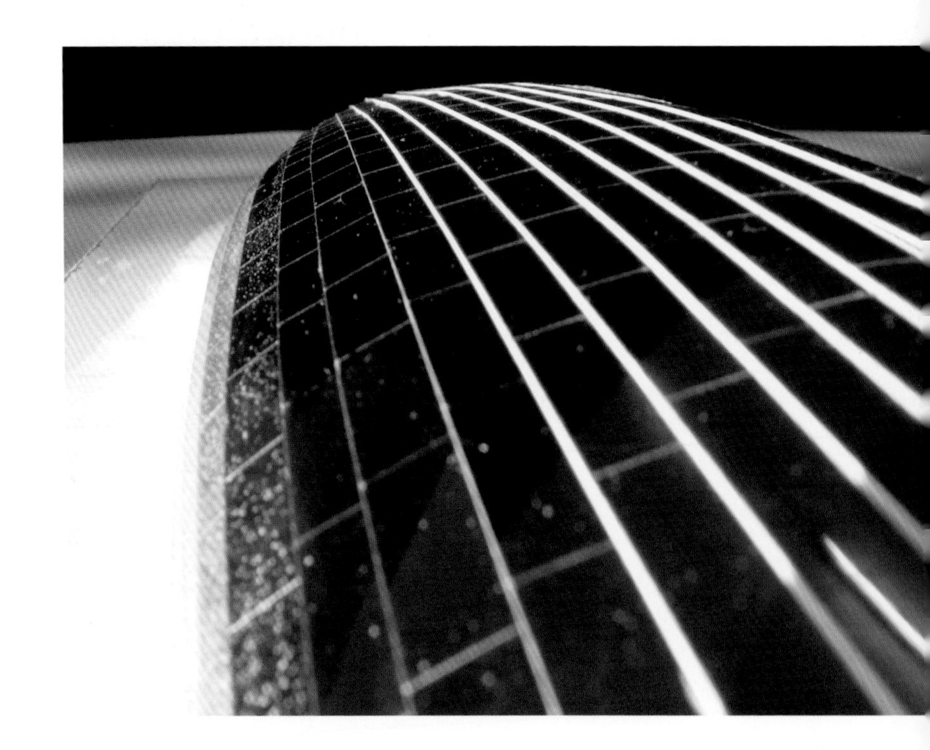

接下来，研究人员制作了一种类似于鱼形的模型，该模型在建筑物主入口以直边切入。这个模型是通过胶合在一起的对开纸板条制成的。将制作好的模型放在玻璃上进行拍摄，从而展现出图书馆倒映在湖中的想法。

In the next stage a study was made of a model somewhat resembling the shape of a fish, cut through with a straight edge on the main entrance side of the building. The model was made by gluing together folio-covered cardboard strips. The completed model was photographed on top of a sheet of glass, thus giving an idea of the reflection of the library in the lake.

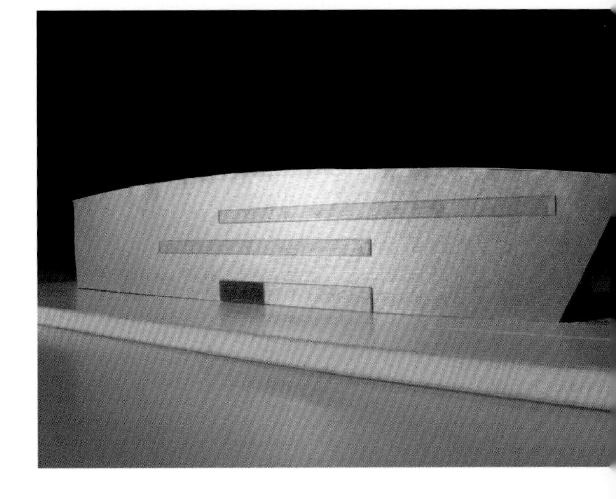

以1 ：50比例构建的模型展现了该图书馆立面及其内部的效果，模型的框架和家具都用桦木胶合板制成。其金属表面是弯曲成型的0.3mm铝板，上面雕刻有线条，代表金属板屋顶。弧形的屋顶横梁，则是通过将多层薄型飞机桦木胶合板黏接到模具上制成的。

由于成本原因，建成的建筑更接近最初草案。

The part-model built in the scale 1 : 50 shows the effect produced by the facade as well as the library interior. Both the model's frame and the fittings were built from birch plywood. The metal surface of the facade is 0.3 mm aluminium sheet bent to shape, into which were inscribed lines representing the metal-sheeting roof. The curved roof beams were made by gluing multiple layers of thin aircraft birch plywood against a mould.

Due to cost reasons, the realised building is closer to the first draft proposal.

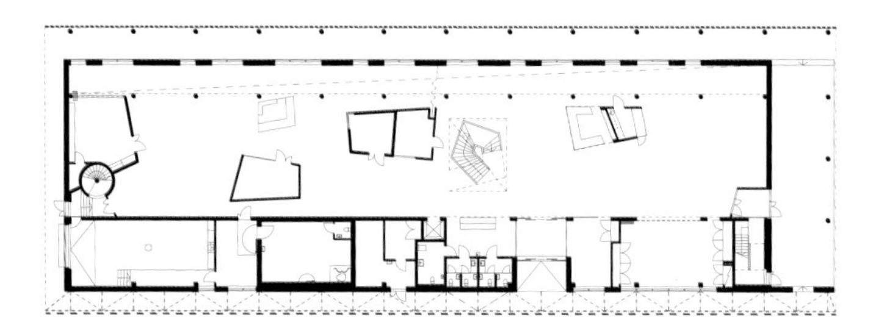

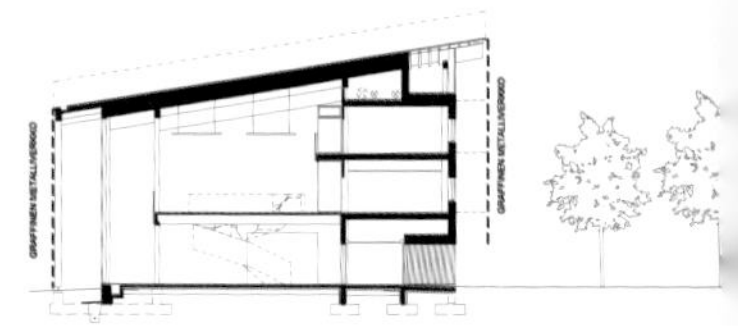

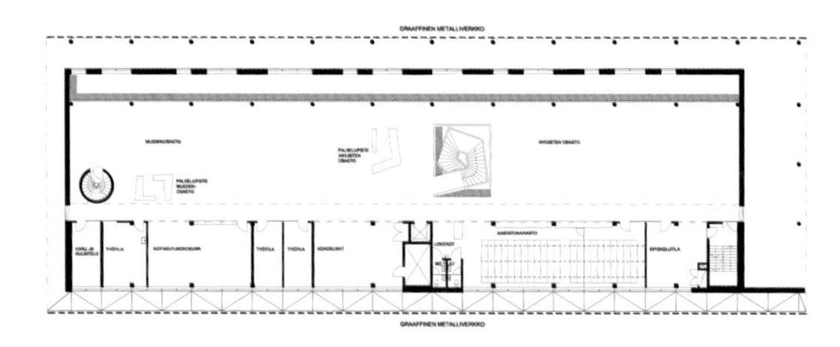
GRAAFFINEN METALLIVERKKO
GRAAFFINEN METALLIVERKKO

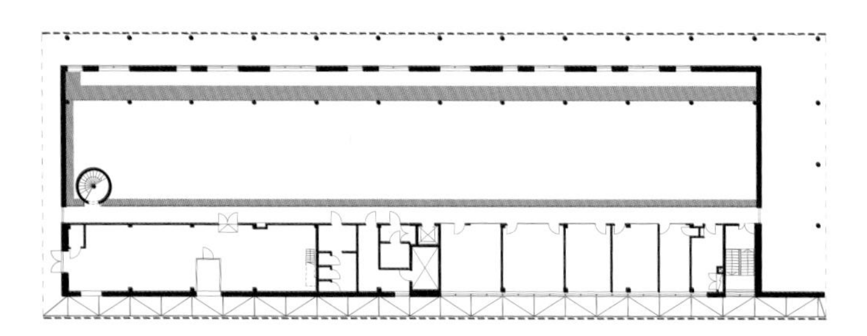

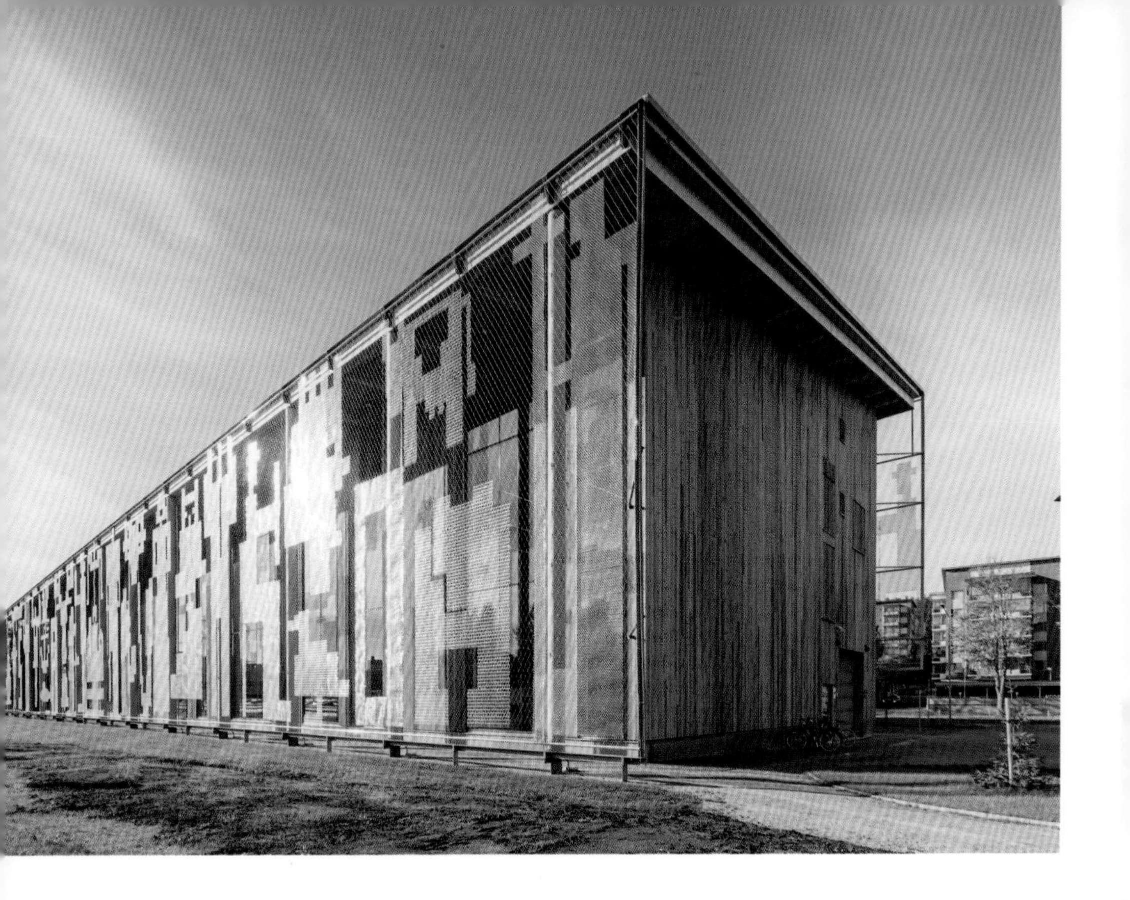

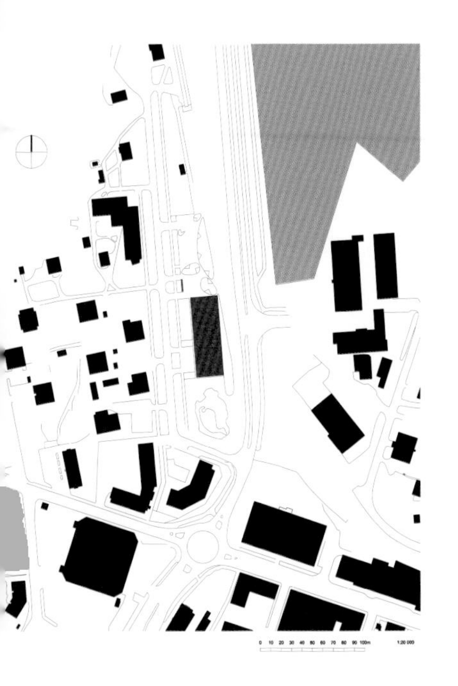

Metal “fishing nets” were strung from the over-hang-ing eaves all the way along the building. Artist Aimo katajamäki drew stlized images on them based on local motifs, such as castles, stemboats, bears, rams’ horns and the unique ringed seals that live in the nearby Lake Saimaa.

屋檐上挂着的金属“渔网”覆盖了整座建筑的外立面。艺术家艾莫·卡塔加梅基（Aimo Katajamaki）根据当地的主题，如城堡、汽船、熊、羊角和塞马湖附近独特的环纹海豹，在建筑外立面上画出了静态的图像。

以图像处理陶瓷
PHOTOSHOP CERAMICS

弗洛拉诺基奥住宅区
芬兰，赫尔辛基，2012年

Flooranaukio Housing
Helsinki, Finland, 2012

这座住宅楼位于前阿拉比亚陶瓷厂附近。将工厂生产过程中产生的废弃瓷器，用于装饰建筑庭院的外墙，让人联想起西班牙建筑师安东尼奥·高迪在巴塞罗那奎尔公园里用破碎陶瓷拼贴的起伏长椅。

The residential building is situated in the vicinity of the former Arabia ceramics factory. Waste porcelain, which is a by-product of the factory's manufacturing process, was used in the yard facade of the building reminding Antonio Gaudi's undulating benches covered with broken ceramics at the Park Güell in Barcelona.

此方案的效果是通过放大阿拉比亚的旧装饰图案，然后应用在预制混凝土上而实现的。第一个比赛阶段的模型是按照 1 ： 100 的比例建造的，陶瓷片被切成装饰图案拼贴成画，黏在弯曲成波浪状的薄纸板上。立面上的开口表示阳台，而公寓的窗户则为一种浅浮雕的形式。

This was achieved by enlarging an old decorative pattern used by Arabia on the surface of the pre-fabricated concrete facade elements. The model for the first competition stage was built in the scale 1 : 100, with a collage of ceramic pieces cut into a decorative pattern glued on top of thin cardboard bent into an undulating form. The balconies are represented as openings in the surface of the facade, while the windows to the apartments are shown as a relief.

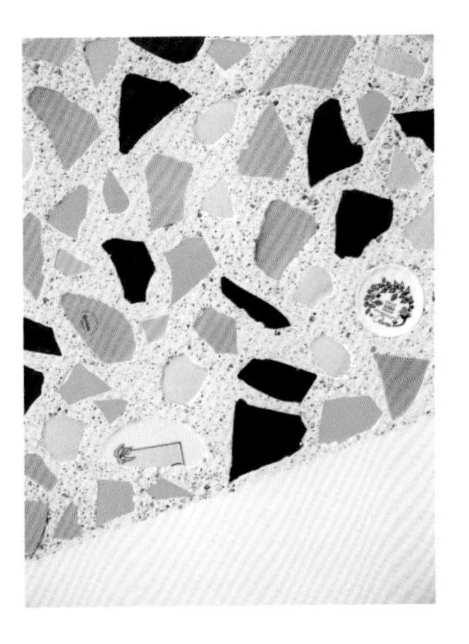

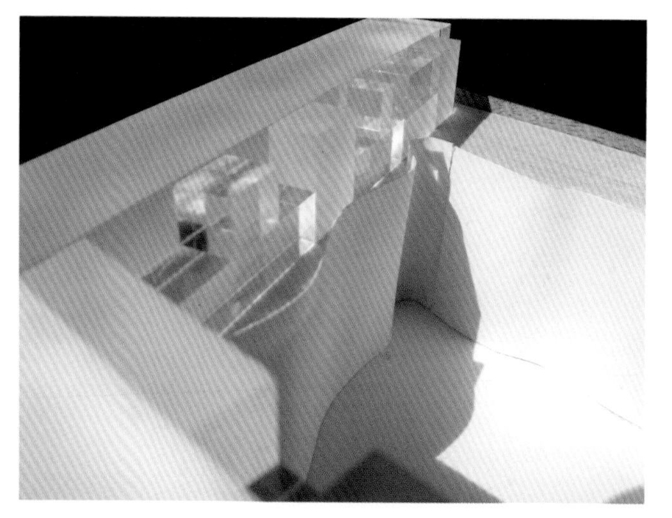
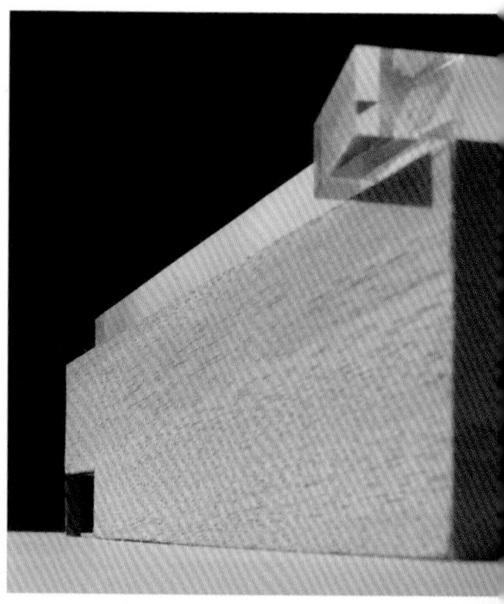

在按照 1 ： 500 比例制作的第二个模型中，我们探索了如何将竞赛阶段产生的想法在最终的设计图上体现。在建筑物屋顶上增加的两层均由亚克力材料制作，或是哑光（住宅），或是亮光（阳台）。该模型的外墙饰面是红木贴面，营造出红砖的效果。

With a second model in the scale 1 : 500, we explored how the concept derived at the competition stage would be placed on the final plot. The two added storeys on the roof of the building are made from acrylic, either with a matt (the dwellings) or high-gloss (the balconies) finish. The model's exterior facades are mahogany veneer, which creates the impression of red brick.

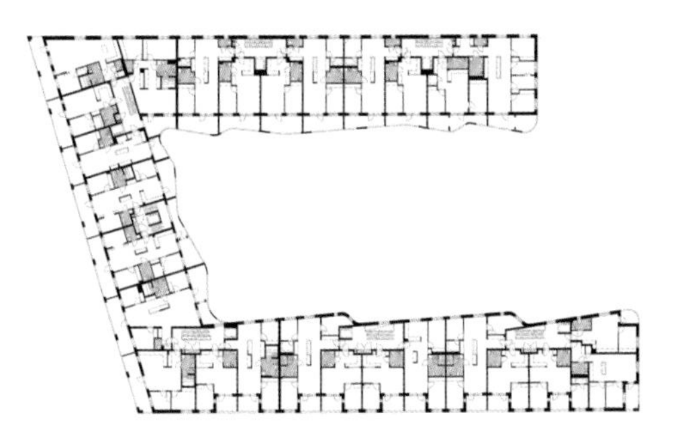

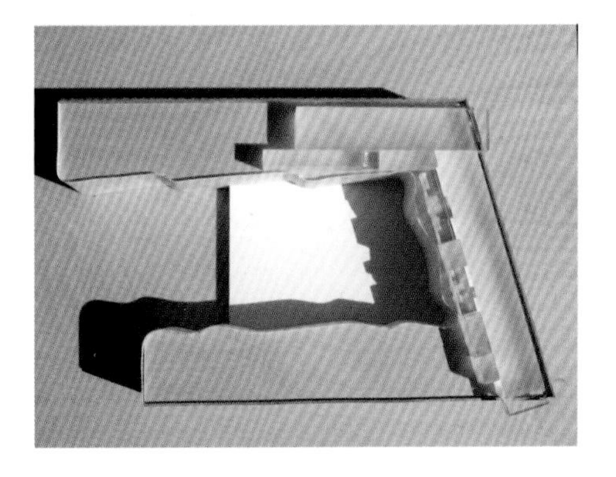

轻金属
LIGHT METAL

林斯托姆服务中心
芬兰，万塔，2011年

Lindström Service Centre
Vantaa, Finland, 2011

这座建筑是为一个从事纺织服务业的公司设计的，拥有大尺度的拉网主立面。在比例模型中，我们探索了网格的形式：网格由易于成形的铝材制成，从不同的方向去看，都以令人惊讶的方式反射光线。模型仅限于表现主立面本身，用来拍照，便于为客户展示。网格表面粘贴着从彩色描图纸上剪下的公司标志。

The main facade of the building for the company in the textile service industry has a pattern of large-scale pulled mesh. The form of the mesh was explored in the scale model: the mesh is aluminium, which is easy to shape, and it reflects light in a surprising way when viewed from different directions. The model was limited just to the main facade itself, which was then photographed for the client presentation. The company's logo cut out from coloured tracing paper was glued on the surface of the mesh.

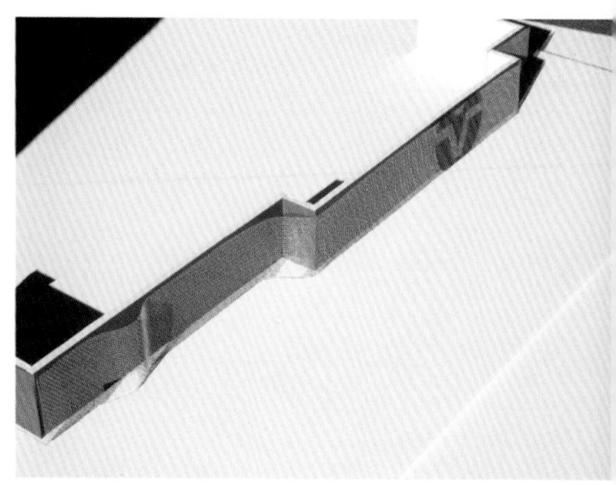

瑟拉彻斯博物馆扩建

芬兰，曼塔，2011年竞赛，二等奖

Serlachius Museum Extension

Mänttä, Finland, 2011 Competition entry, 2nd Prize

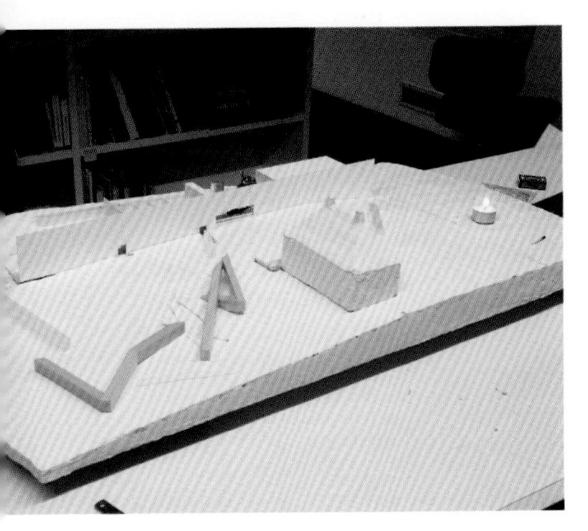

在竞赛方案中，展厅和公共空间被“包裹”成一个木质容器，放置于老庄园的附近，周围环绕着公园。楼层之间的流通路线从玻璃通道延伸到公园。在草案设计阶段，我们制作了一个石膏地形模型，嵌入了由白色硬纸板制成的建筑物。那些像是章鱼触角的通道，是用半透明的树脂浇铸而成的。

然而，最终的地形模型是由能更准确体现结构的白色硬纸板制成的，其照片展现在竞赛展示板中。

In the competition proposal, the exhibition halls and public spaces have been “packaged” into a wood-clad container placed adjacent to the old manor house and surrounded by park. The circulation routes between the levels occur via glazed passages that project out into the park. At the sketch design stage, a plaster terrain model was prepared, into which the building, made from white cardboard, was embedded. The passages, somewhat resembling octopus tentacles, were cast in semi-transparent resin.

The final terrain model, however, was built more precisely from white cardboard, the photos of which were included in the competition display boards.

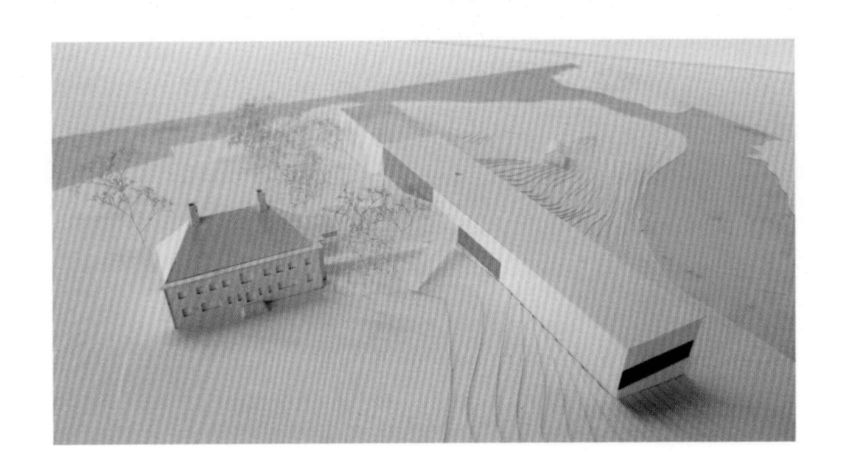

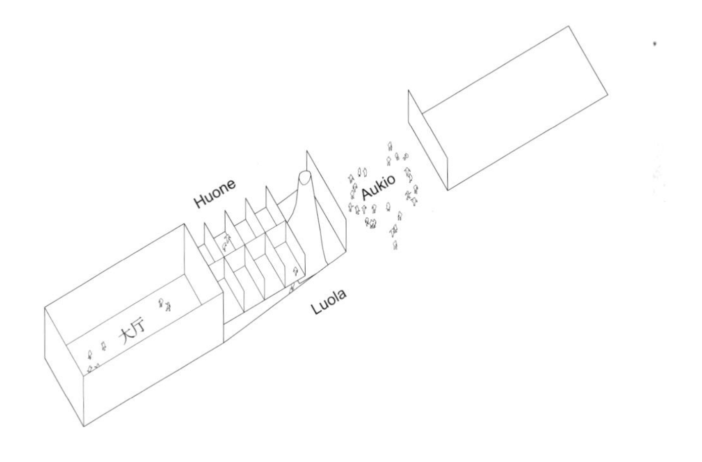
Huone
Aukio
Luola
大厅

空间中的空间
SPACE IN A SPACE

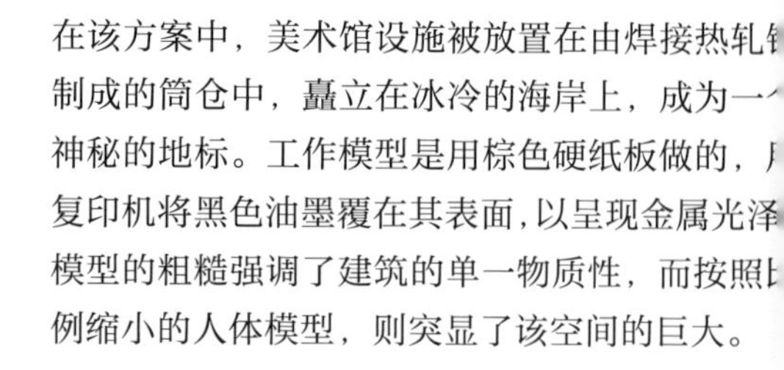

在该方案中，美术馆设施被放置在由焊接热轧
制成的筒仓中，矗立在冰冷的海岸上，成为一
神秘的地标。工作模型是用棕色硬纸板做的，
复印机将黑色油墨覆在其表面，以呈现金属光泽
模型的粗糙强调了建筑的单一物质性，而按照
例缩小的人体模型，则突显了该空间的巨大。

格陵兰国家美术馆
格陵兰，努克，2010年竞赛

National Gallery of Greenland

Nuuk, Greenland, 2010 Competition entry

In the proposal the art museum facilities are place
in a silo built from welded hot-rolled steel, whic
rises as a mysterious landmark on the shore of th
ice sea. The working model is made of brown car
board, onto the surface of which black ink has bee
spread using a photocopier in order to give the im
pression of a metallic patina. The coarseness of th
model emphasises the architecture's mono-mate
riality, while the small-scale figures emphasise th
monumentality of the space.

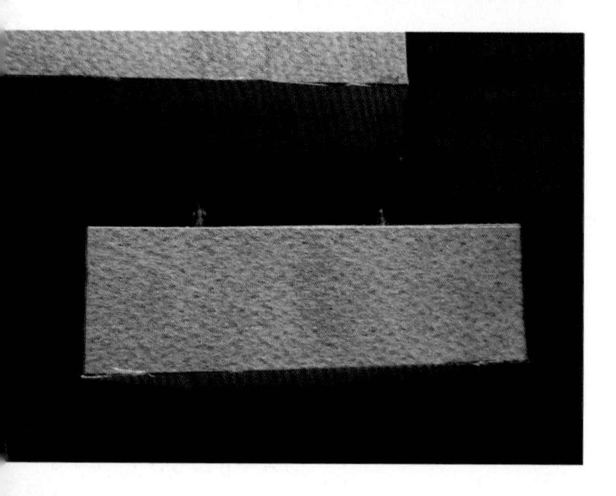

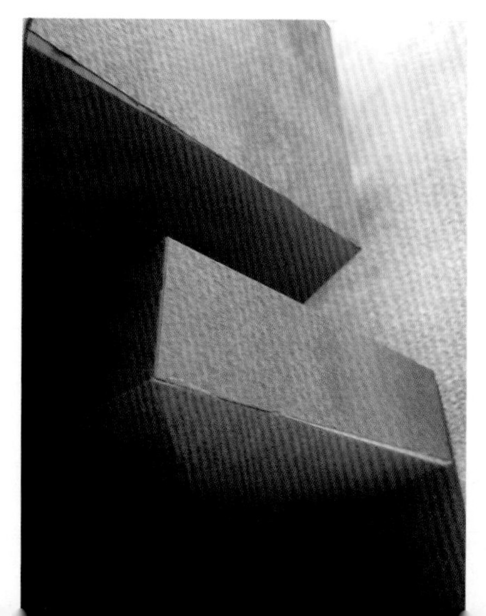

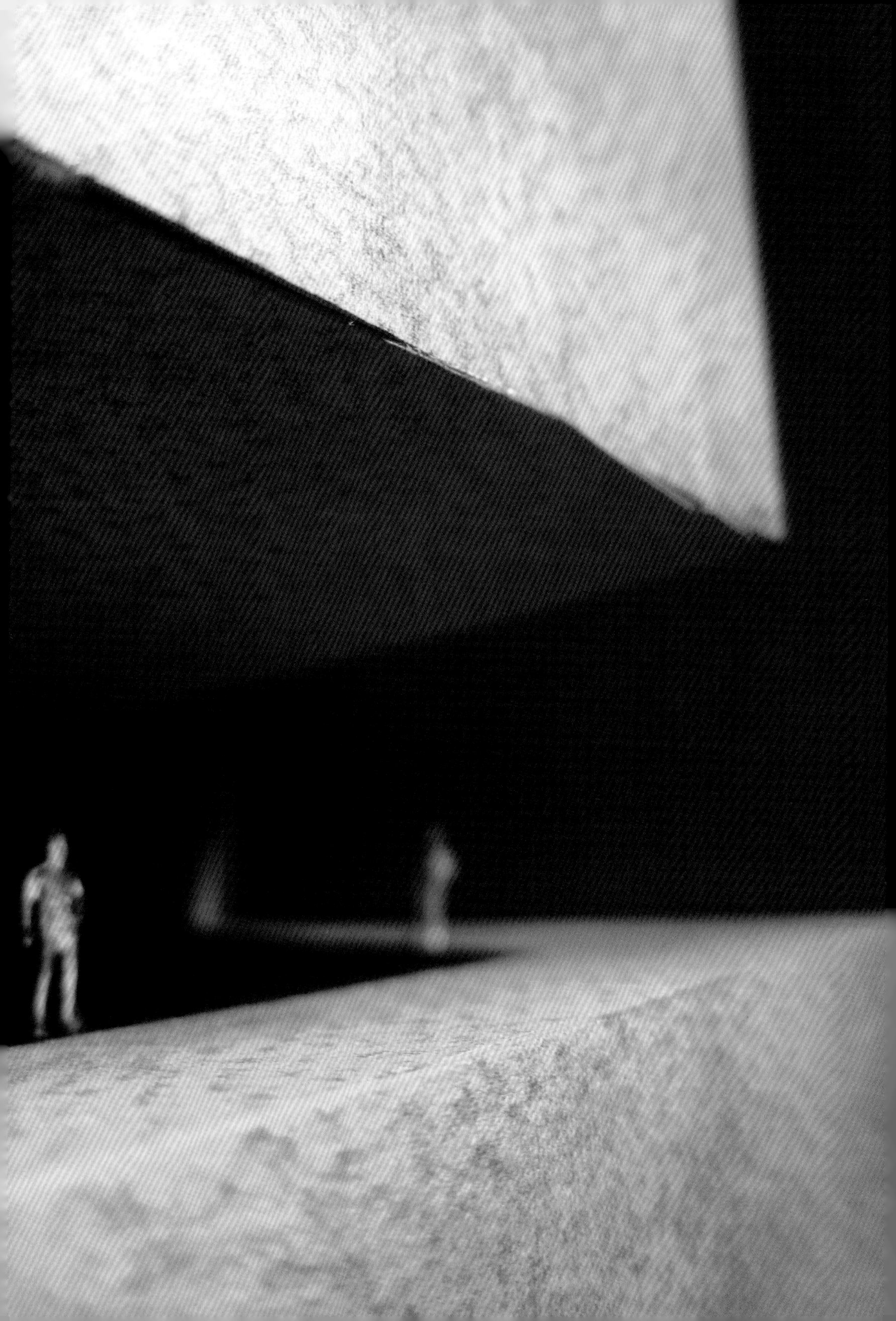

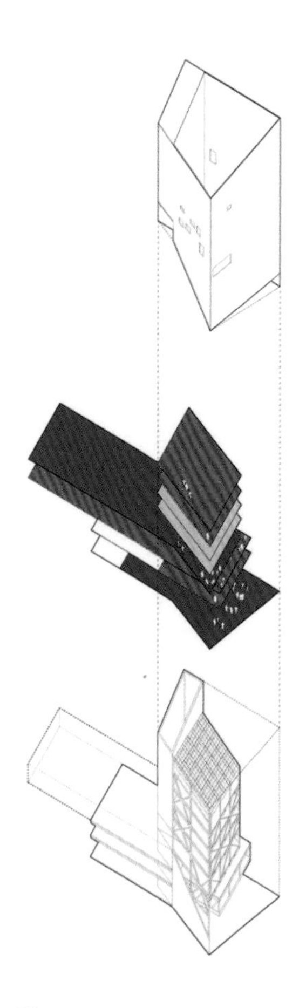

以 1 ： 500 比例制作的现场模型，用于探索美术馆在景观中的位置，没有树木的海岸线上充斥着巨大的混凝土住宅。该模型由白色硬纸板做成，地形的层高线足够密集，岩石海岸线的有机可塑性与人造建筑形成鲜明的对比。

The position of the museum in the landscape was explored with a site model built in the scale 1 ： 500. The treeless shoreline is dominated by enormous concrete housing colossi. The model is built from white cardboard. The height lines of the terrain are sufficiently dense that the organic plasticity of the rocky shoreline would stand out in contrast to the man-made architecture.

从树桩到概念

FROM A STUMP TO A CONCEPT

上海世博会芬兰馆

中国，上海，2008年竞赛，佳作奖

Shanghai Expo 2010 Finnish Pavilion

Shanghai, China, 2008 Competition entry, Purchase

竞赛方案的想法源于一个树桩。概念模型来自从一棵倒下的树干上锯下的圆盘切面，从中雕刻出一个空腔，空腔内按一层设计布局的草案布置。腔体的立面覆以镜面材料。建筑模型的目的是为了展现有机木材表面和直线切割后的对比效果。

The idea for the competition proposal came from a tree stump. The conceptual model was made from a disk sawn from the trunk of a fallen tree, from which was carved a cavity that followed the draft design layout for the ground floor. The facade surfaces of the cavity were covered with mirrored contact plastic. The objective of the architecture was the contrast between the organic wood surface and rectilinear structure cutting through it.

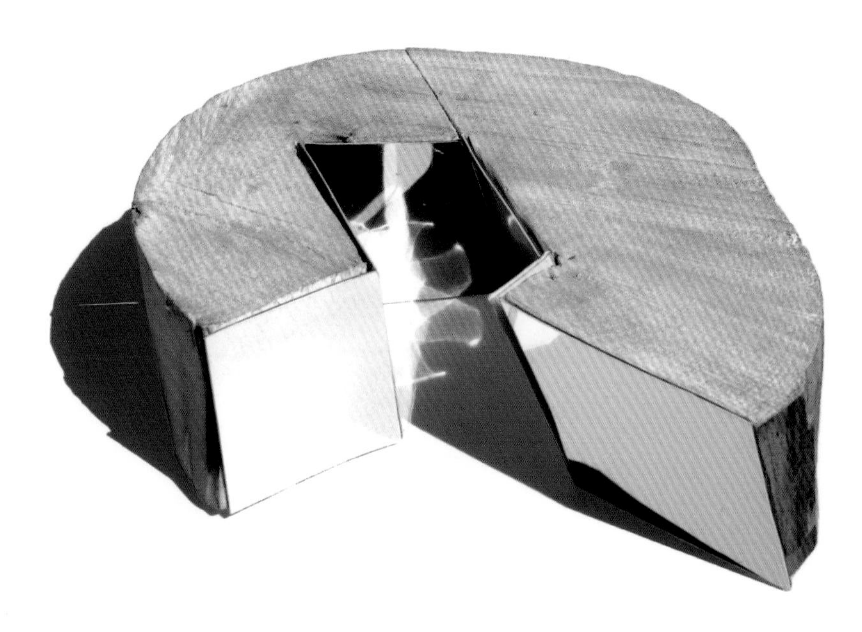

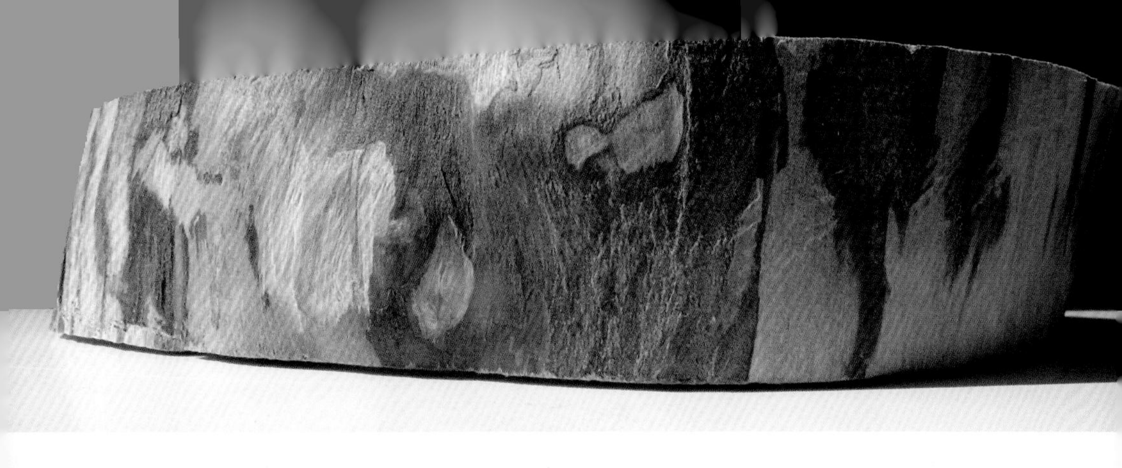

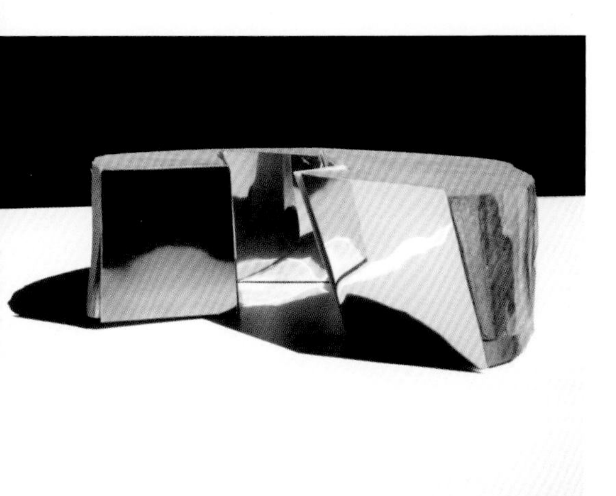

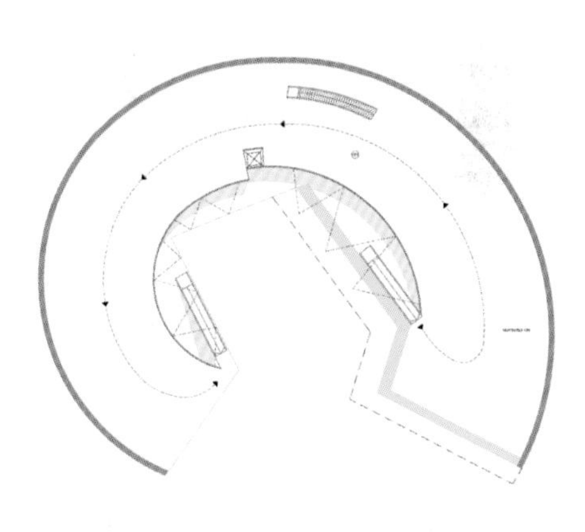

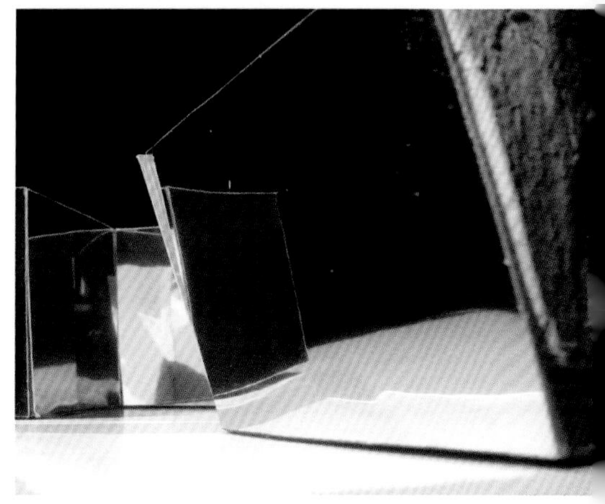

另一个更精确的模型是用胶合制作的，比例为1 ：200。外墙的弯曲形状是通过将薄胶合板黏合在一起，并将其压入模型底板上预先切割出的所需形状的凹槽而实现的。在内部庭院的玻璃后面，可以看到展示北极光的墙面。墙面由投影仪放出的幻灯片覆盖，图像提前被影印到幻灯片上。竞赛方案的图片和立面图纸直接来自模型的照片。

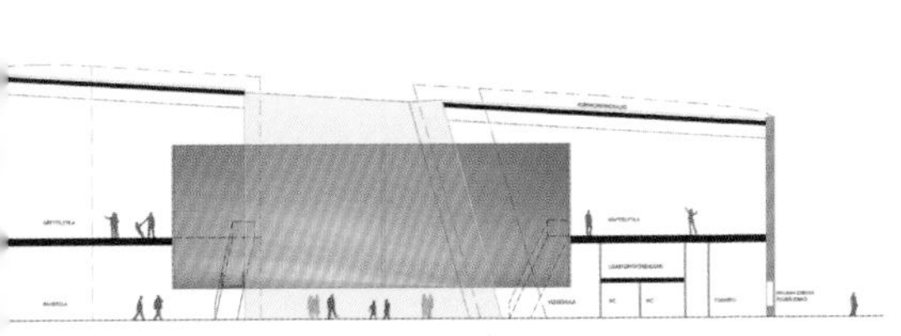

Another more precise model was built in plywood at the scale 1 : 200. The curved form of the outer wall was achieved by gluing together thin plywood veneers and forcing them into a groove of the desired shape cut into the baseboard of the model. The wall displaying the Northern Lights, which can be seen behind the glass in the internal courtyard, was made from overhead-projector transparencies, on to which the image was photocopied. The illustrations and facade drawings for the competition proposal were adapted directly from photos of the model.

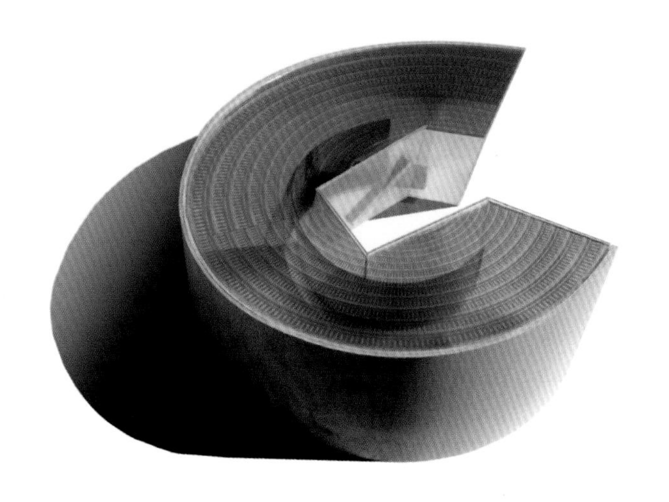

该模型的屋顶结构代表太阳能板。将用
Photoshop（图像处理软件）处理后的
计算机微电路图像打印在透明纸上，再
压在两块亚克力板之间。

The model's roof structure represents solar panels. A transparent print of an image edited in Photoshop depicting a computer micro-circuit was pressed between two acrylic sheets.

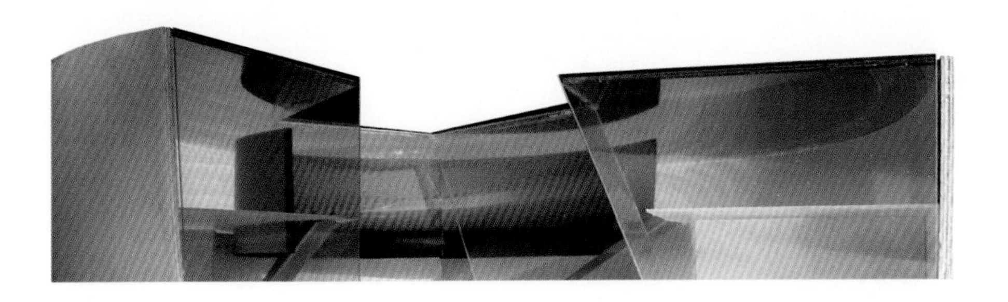

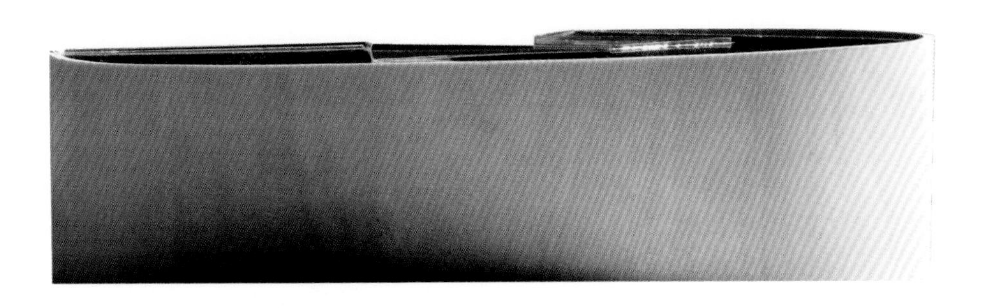

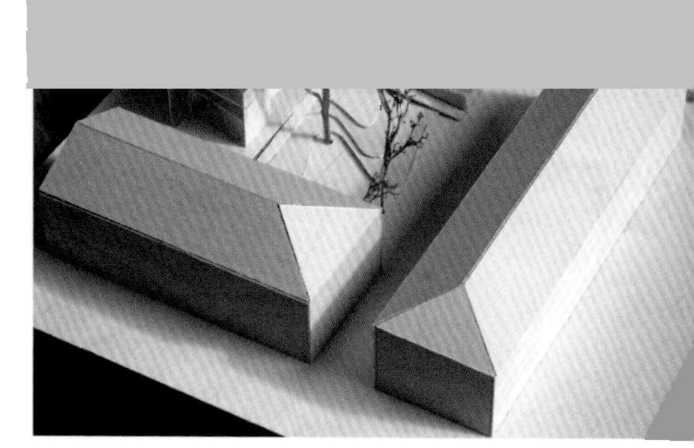

光与影
IGHT AND SHADOW

弗里达别墅
芬兰，波尔沃，2010年

Villa Frida
Porvoo, Finland, 2010

两户人家的住所建在一个有历史背景的木屋中。该概念模型是用薄胶合板和亚克力板制成的，目的是向客户展示如何在基地上建造这个带有地下车库和坡道的、相对较大的新建筑，以保留庭院有多棵大树的氛围。

Dwellings for two families were to be built in a historical milieu of wooden houses. The concept model was built from thin plywood and acrylic sheets in order to show the clients how the relatively large new building, with its underground garage and ramps, could be placed on the plot, so that the ambience of the yard, with its large trees, could be preserved.

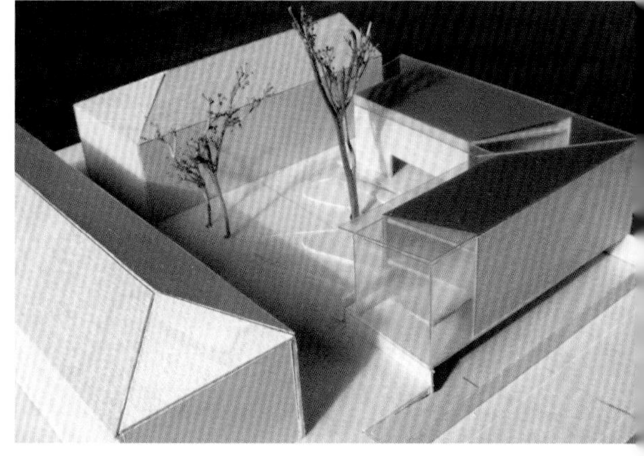

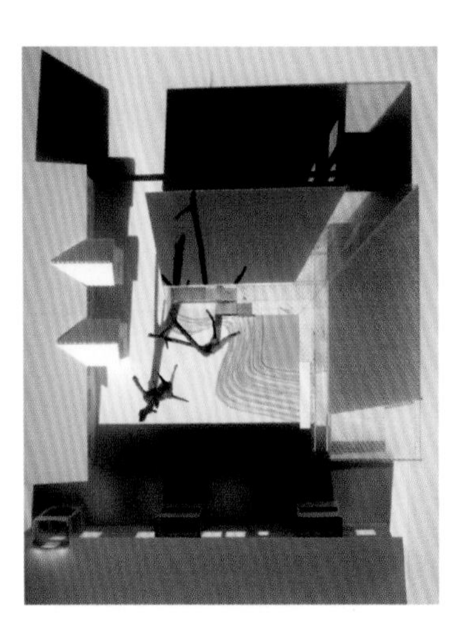

按照 1 ： 100 比例制作出的更详细的模型，用于市政府的检查
和邻里协商。该模型由白色硬纸板制成，其表面黏有周围建筑
立面的照片，目的是通过照片来展示项目，因此照明必须尽可
能真实。剪下来的代表树干的树枝，在照片中帮助模型创造出
城市老木屋的氛围。

A more detailed model was made in the scale 1 : 100 for the in
spection of the city authorities and neighbourhood consultations
The model was built from white cardboard, on the surface of whic
were glued photocopy prints of the facades of the surroundin
buildings. The objective was to present the project through photo
graphs, and thus the lighting had to be as realistic as possible. Cu
twigs representing tree trunks helped in creating in the photos th
ambience of the urban neighbourhood of old wooden houses.

三维平面图

3D FLOOR PLAN

海门林纳省档案馆

芬兰，海门林纳，2009年

Hämeenlina Provincial Archive

Hämeenlinna, Finland, 2009

此建筑由一个封闭的档案室、办公楼和公共空间组成。建筑模型展示了档案馆开放的临街楼层，除了一个公众服务点以外，还包括研究人员工作站、一个图书馆和一个小型演讲厅。按比例制作的模型由胶合板、投影仪幻灯片和彩色纸板制成。在这个阶段，模型已经提供了未来建筑的配色方案。

The building consists of an enclosed archives container, office wing and public spaces. This model showed the open street level floor of the archives, which includes, in addition to a service point for the general public, researcher workstations, a library and a small lecture hall. The scale model was built form plywood, overhead-projector film and coloured cardboard. Already at this stage, the model gave an indication of the future building's colour scheme.

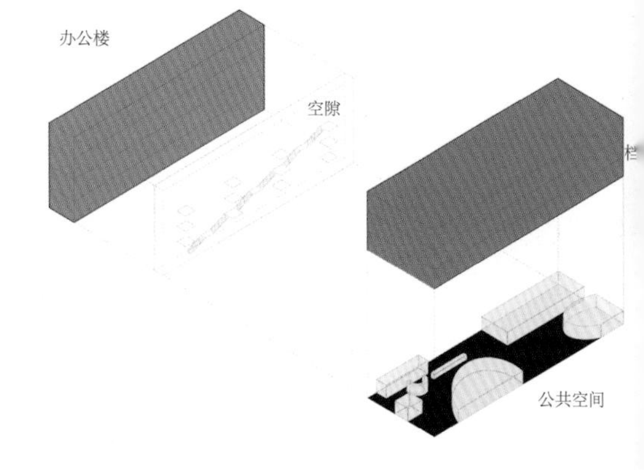

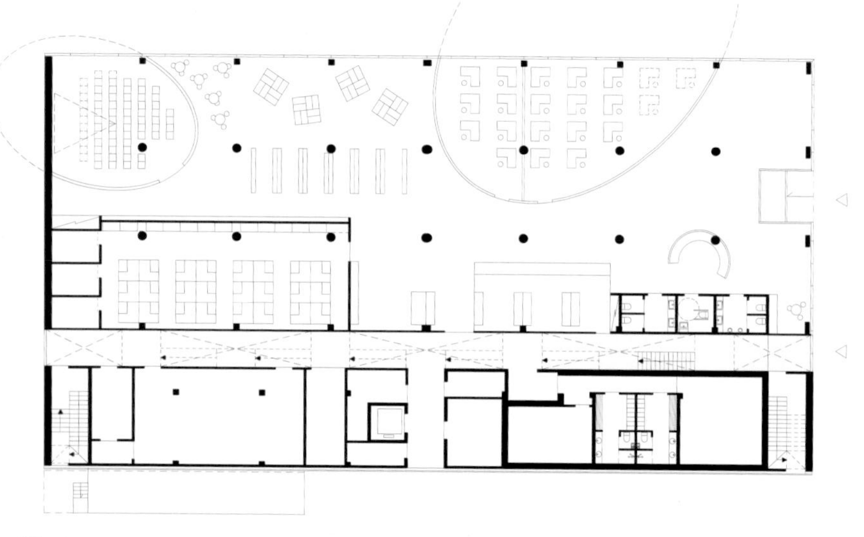

错综复杂的清晰

LABYRINTHIAN CLARITY

1 乔凡尼·巴蒂斯塔·皮拉内西（Giovanni Battista Piranesi，1720–1778年），意大利雕刻家和建筑师。他以蚀刻和雕刻现代罗马以及古代遗迹而成名。他的作品于1745年首次出版，并在他去世后被反复印刷。强烈的光、影和空间对比，以及对细节的准确描绘，是其作品的特点。在《卡里·德·英芬辛内》（*Carceri d'Invenzione*）中有许多极富想象力的监狱场景，是他最具创造性的作品。
——译者注

微观医学实验室中心

芬兰，图尔库，2009年竞赛

Micromedicum Laboratory Centre

Turku, Finland, 2009 Competition entry

在方案中，实验室建筑群分为三个区域，中间区域是一个高大的三角形大堂空间，由楼梯和桥梁将建筑物的不同部分连接在一起。比例为1 ：200的模型展示了中央大厅皮拉内西（Piranesi[1]）式效果及其与翼楼绿色中庭庭院的连接。中庭内的树木由特别适用于建筑模型的可塑材料制成。

In the proposal the laboratory complex is divided into three blocks, at the centre of which is a tall triangular lobby space, with stairs and bridges linking together the different parts of the building. The 1 : 200 model shows the Piranesi-like atmosphere of the central lobby and its connection to the green atrium courtyard of the wing buildings. The trees in the atrium are made from mouldable mass intended especially for model building.

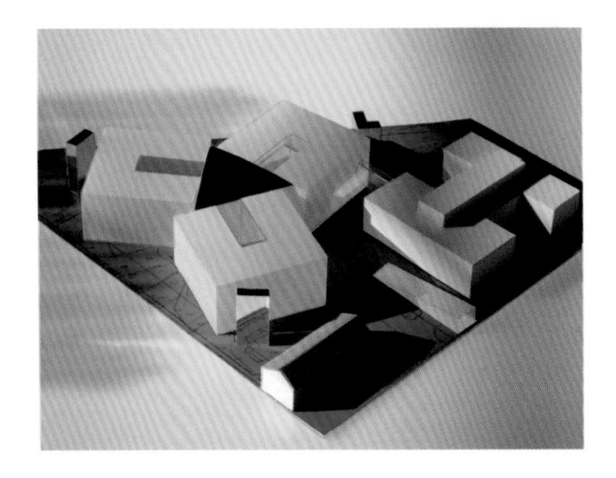

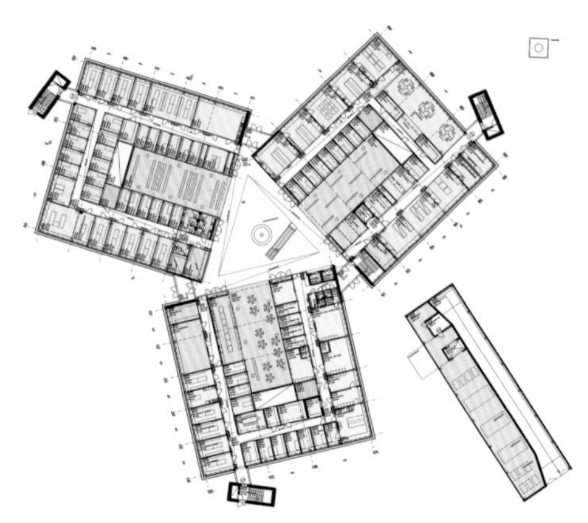

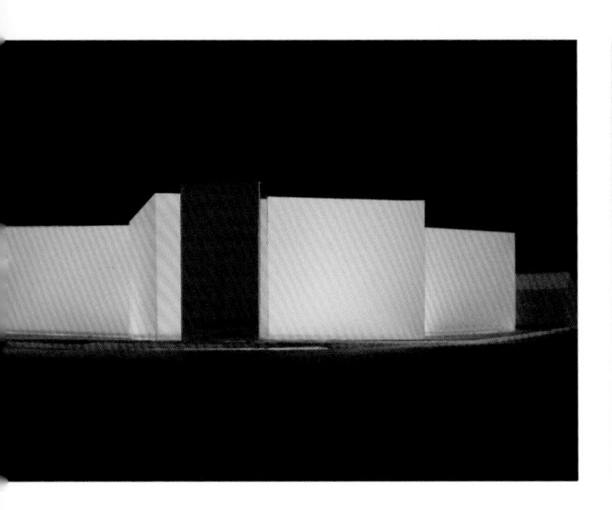

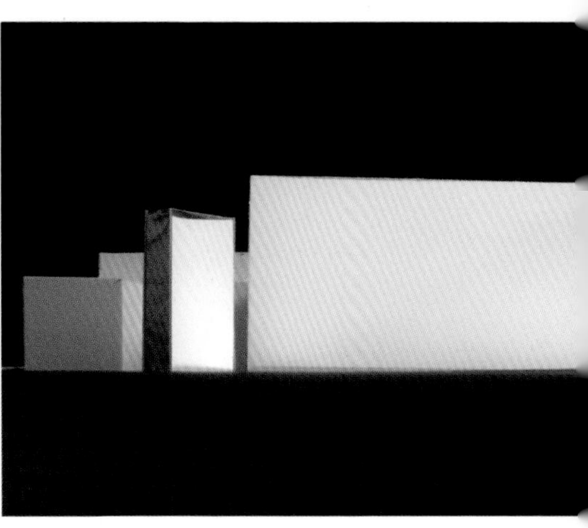

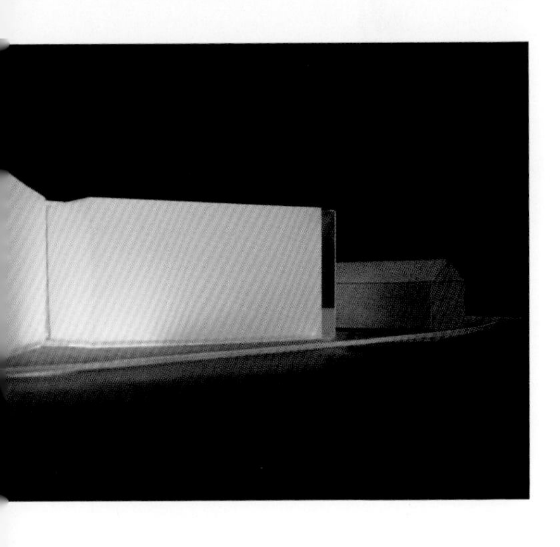

比例为 1 ： 500 的总平面模型是用描图纸折叠出建筑物的形状，并将其固定在绿色纸板制作的总平面复制图上。通过在每个照明不足的翼楼下面放置不同颜色的透明板，展现出夜间建筑外墙的变化。在拍照时，很容易改变外墙的颜色和亮度。

The site plan model in the scale 1 : 500 was made by folding from tracing paper the forms of the buildings, and fixing them to a site plan photocopied on green cardboard. The architectural character of the facades, which fluctuates at night-time, was shown by placing different coloured transparent sheets beneath each of the under-lit building wings. During the photo session it was thus easy to alter the colour and its intensity.

堆积如山的地板

A MOUNTAIN OF FLOORS

阿尔玛媒体大厦

芬兰，赫尔辛基，2009年

Alma Media Building

Helsinki, Finland, 2009

该模型是一个快速到“一晚完工”的草案模型，在一个具有挑战性的场地上，显示出大型媒体公司的设计布局。用户需要大量的开放式楼层。在设计中，楼层面积随着楼层上升而减少，并且随着场地和周围建筑物的方向螺旋式堆叠在彼此的顶部。代表楼层的亚克力板由垂直连接的木销贯穿，外立面的实体区和开放区由粘贴在亚克力板边缘的白色条带区分，上面贴有公司各种产品品牌的贴纸。

This model was a quick "one night" sketch model showing the placement of the large media house on a challenging plot. The user wanted extensive open plan floors. In the design, the levels decrease in floor area as one moves upwards and are stacked on top of each other in a helix formation following the orientations of the plot and surrounding buildings. The acrylic boards representing the floors are pierced by wooden dowels representing the vertical connections. The solid areas and openings in the facades are differentiated by white strips glued on to the edges of the acrylic boards, onto which are attached stickers of the company's various product brands.

mikko.fi
Lapin Kansa
ETUOVI
bo vision.se
Lapin Kansa

Lapin Kansa
ETUOVI
POHJOLAN SANOMAT

POHJOLAN SANOMAT
SATAKUNNAN KANSA
mikko.fi
Lapin Kansa
ETUOVI
mikko.fi
bo vision.se

POHJOLAN SANOMAT
SATAKUNNAN KANSA
ILTALEHTI
mikko.fi
Lapin Kansa
ETUOVI
bc-vision.se

ILTALEHTI
bc-vision.se
ETUOVI
Lapin Kansa
SATAKUNNAN KANSA
POHJOLAN SANOMAT
KAINUUN SANOMAT
mikko.fi
AAMULEHTI

透明颜色

TRANSPARENT COLORS

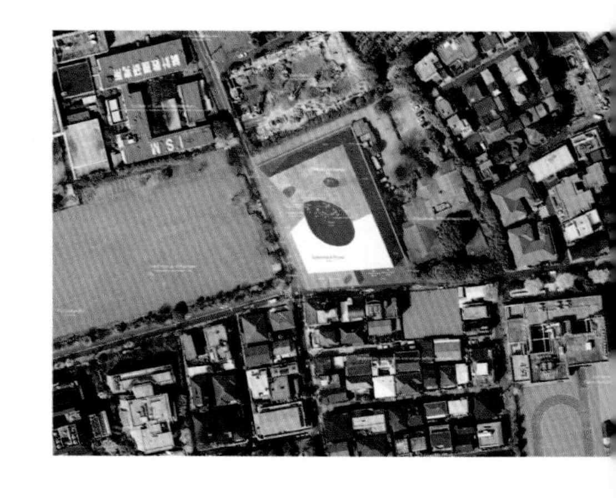

芬兰驻日本东京大使馆

日本，东京，2009年竞赛

Embassy of Finland in Tokyo

Tokyo, Japan, 2009 Competition entry

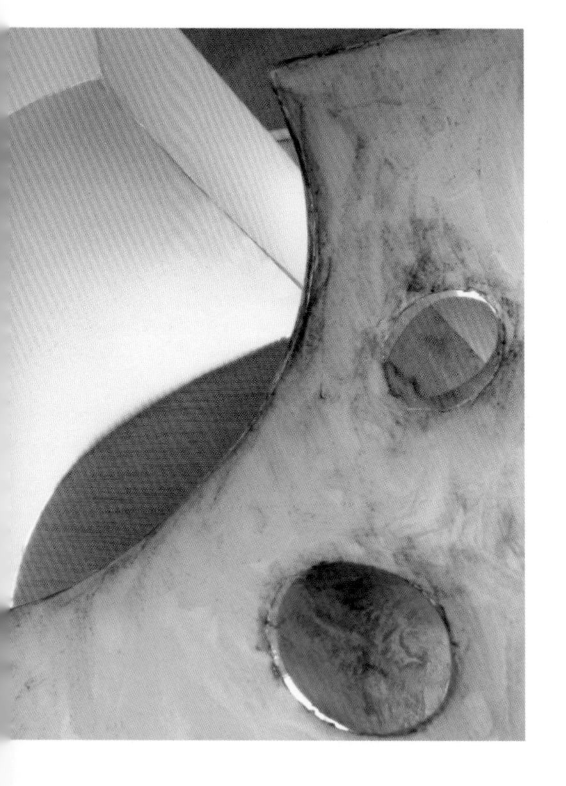

一座相当大的住宅建筑放置在与新使馆大楼相同的狭窄场地上。在我们的方案中，这些功能共同形成一个类似于划艇浆的统一体。在住宅楼的中心建造了一个共享入口的庭院，环绕着睡莲池。按照 1 ： 500 的比例从描图纸上剪下不同的替代物，并从下方照亮，使馆一侧用水彩绘画来凸显，微微发亮的白色住宅区与种植竹子的绿色大使馆区相得益彰。

A fairly large residential building was to be placed on the same constricted plot as the new embassy building. In our proposal, the functions together form a uniform volume that resembles the oarlock of a rowing boat. A shared entrance courtyard was created in the centre of the complex, around a water lily pond. Varying alternatives for the massing were cut from tracing paper in the scale 1 : 500 and lit from below. The embassy side of the complex was highlighted by painting it with water colours. The shimmery white residential section created a counterpart to the green, partly bamboo-clad embassy section.

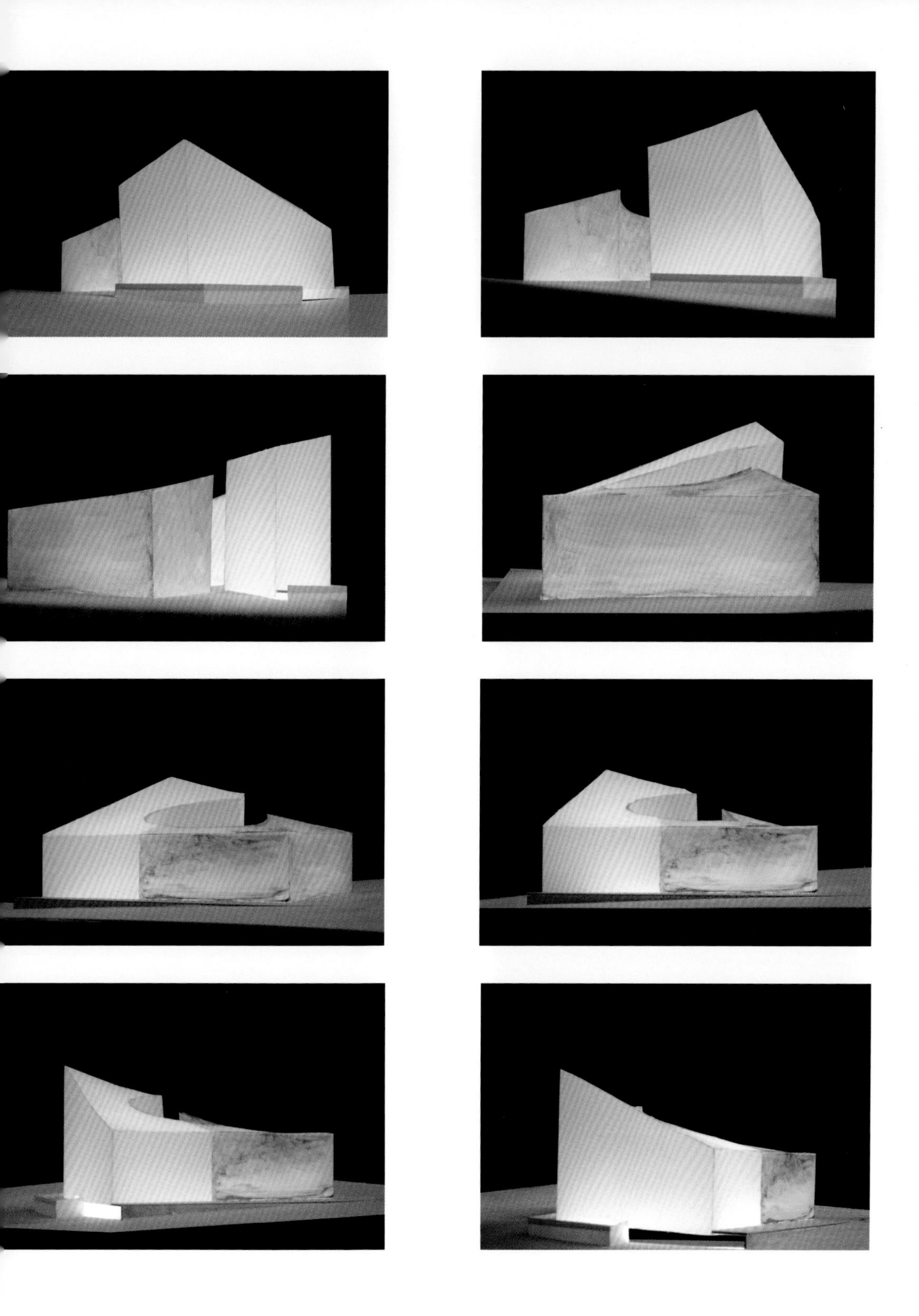

最终的模型由亚克力板制作，之后拍摄图片，用于竞赛展示板上。庭院里的睡莲池塘做成另一个详细的模型，庭院中矗立着竹子，绿色“吸管”插入表面覆盖有反射薄膜的地基中，并剪切睡莲放置在薄膜表面。

The final model was ground from a solid acryli piece and photographed to provide images for th competition display boards. A separate detaile model was made of the lily pond in the courtyard from which rose bamboo stems. Green drinkin straws were inserted into the base which wa covered with a reflective film, and the water lilie were cut into the surface of the film.

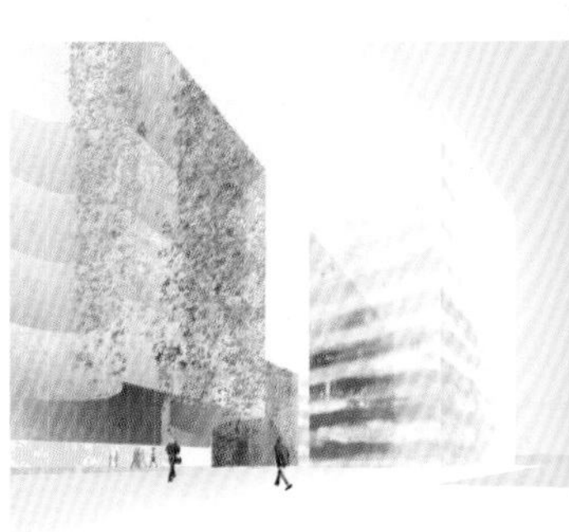

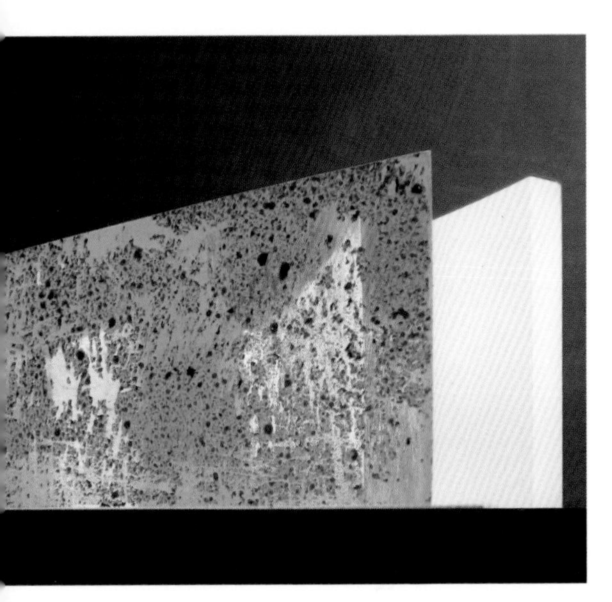

在电影中打印

PRINTS ON FILM

卡雷利安民乐学院（考帕）

芬兰，库恩，2008

Academy for Karelian Folk Music Koppa

Kuhmo, Finland, 2008

这座小木屋包含了民乐表演和研究的所有设备。按比例制作的模型由松木条、胶合板、亚克力和白色硬纸板制成。模型的屋顶可拆卸，这样可以看到内部的平面布局。建筑的纵向外立面覆有金属丝网，上面编织了卡雷利安装饰图案。在模型中，图案由幻灯片呈现。这些图案根据光线的方向，投射成建筑外墙上的阴影效果。

The small wooden building will contain facilitie for both the performance and research of folk mu sic. The scale model is made from pine battens plywood, acrylic and white cardboard. The roof c the model is detachable, so that the floor plan ca be viewed. The longitudinal facades of the build ing are clad with metal mesh, into which Karelia decorative motifs have been woven. In the mode this is represented by overhead projector transpar encies onto which the images have been printed Depending on the direction of the light, the pat terns are replicated as shadows on the walls of th building.

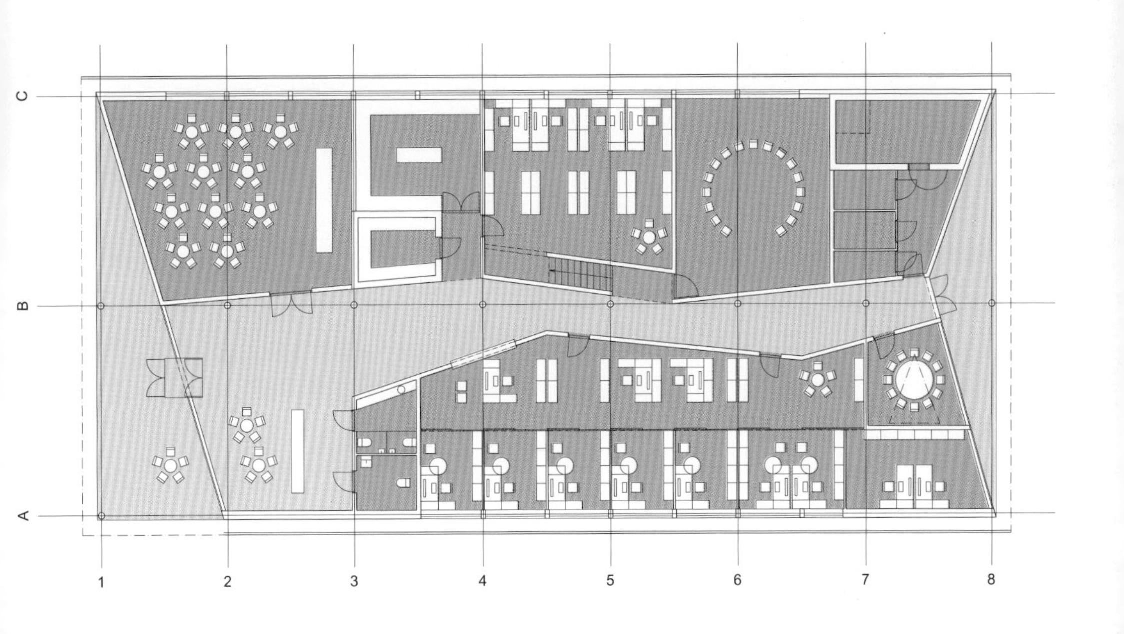
C
B
A
1
2
3
4
5
6
7
8

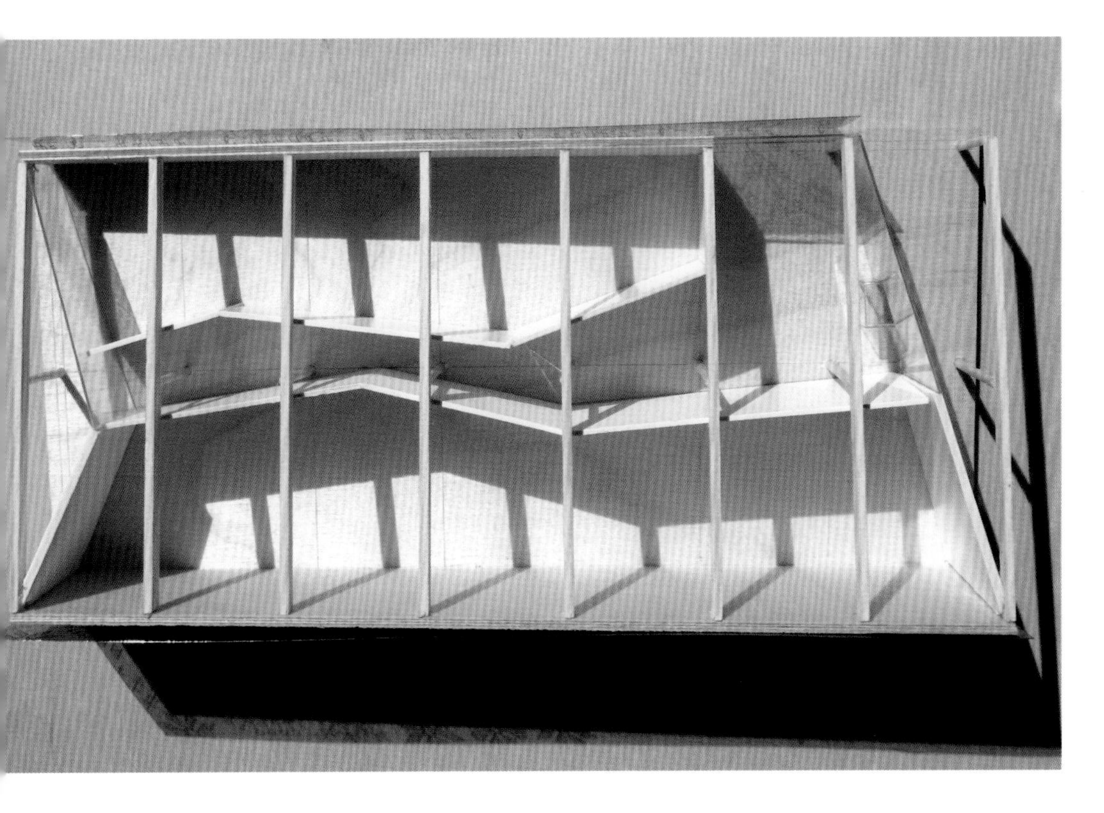

有光泽的地板

CARDBOARD WITH PATINA

塔利——伊汉达拉战争博物馆

芬兰，拉彭兰塔，2008年竞赛

Tali-Ihantala War Museum

Lappeenranta, Finland, 2008 Competition entry

在建筑竞赛方案中，这个小型博物馆展现了芬兰与苏联的决战历史，给人留下深刻的印象，彷佛是战斗结束后部分沉入地下的掩体或残骸。建筑外墙是耐候钢[1]材料，在上面蚀刻或雕刻着军队番号和他们的行动轨迹。

在模型中，将图像复制到棕色硬纸板上，并在纸板上切下箭头来表示部队的行动路线。类似战术地图的透明的影印图像粘贴在建筑入口立面的亚克力板上。

[1] 耐候钢最早起源于考顿钢（Corten Steel），广泛用于火车车厢、集装箱及桥梁的制作。耐候钢被用作建筑立面材料，在北美、西欧、澳大利亚及亚洲的日本、韩国都有一定的使用历史。最早的例子是1965年建成的芝加哥市民中心大厦，采用了暴露的耐候钢结构体，以及耐候钢板和玻璃的混合幕墙。——译者注

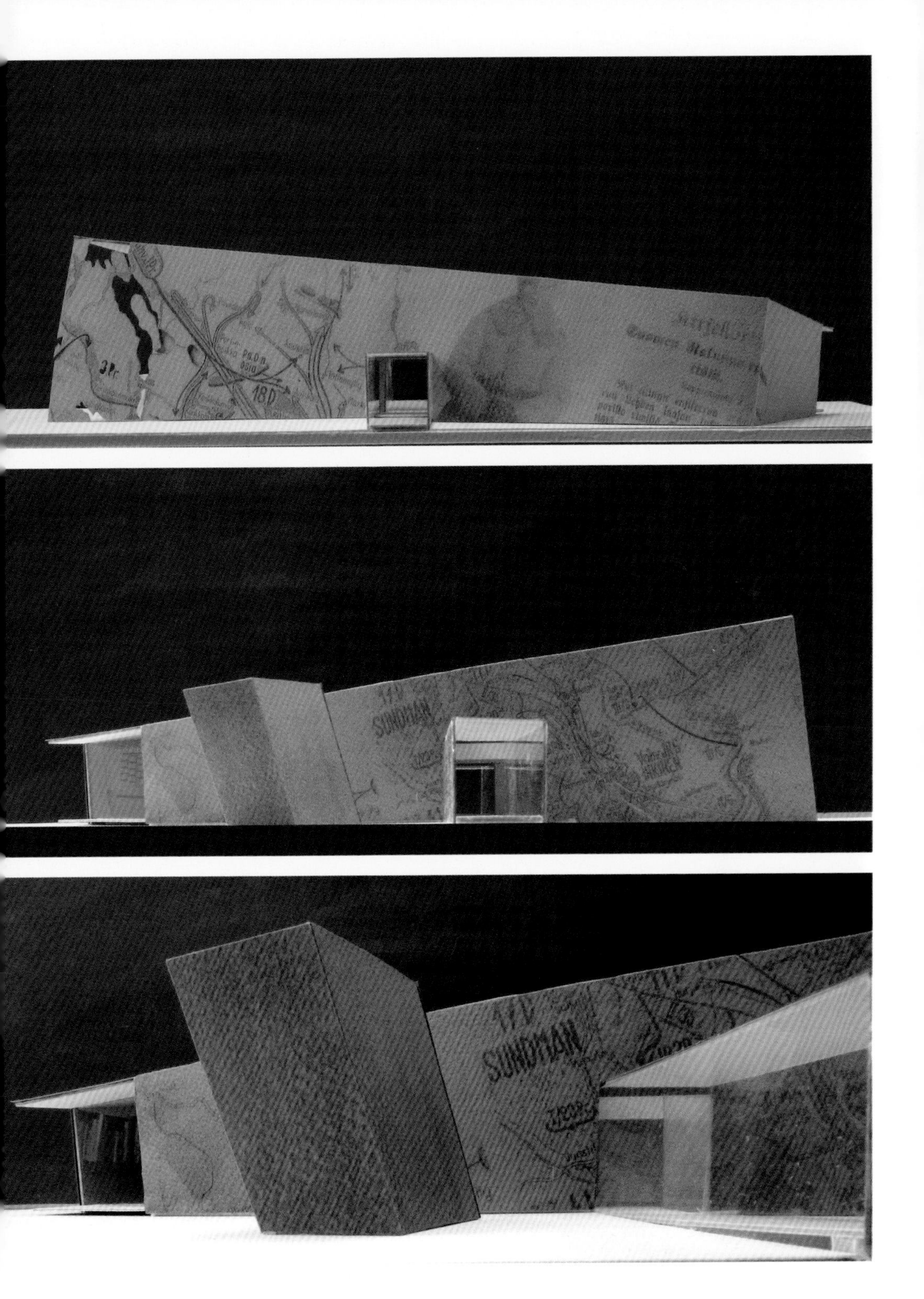
18D
SUNDHAN

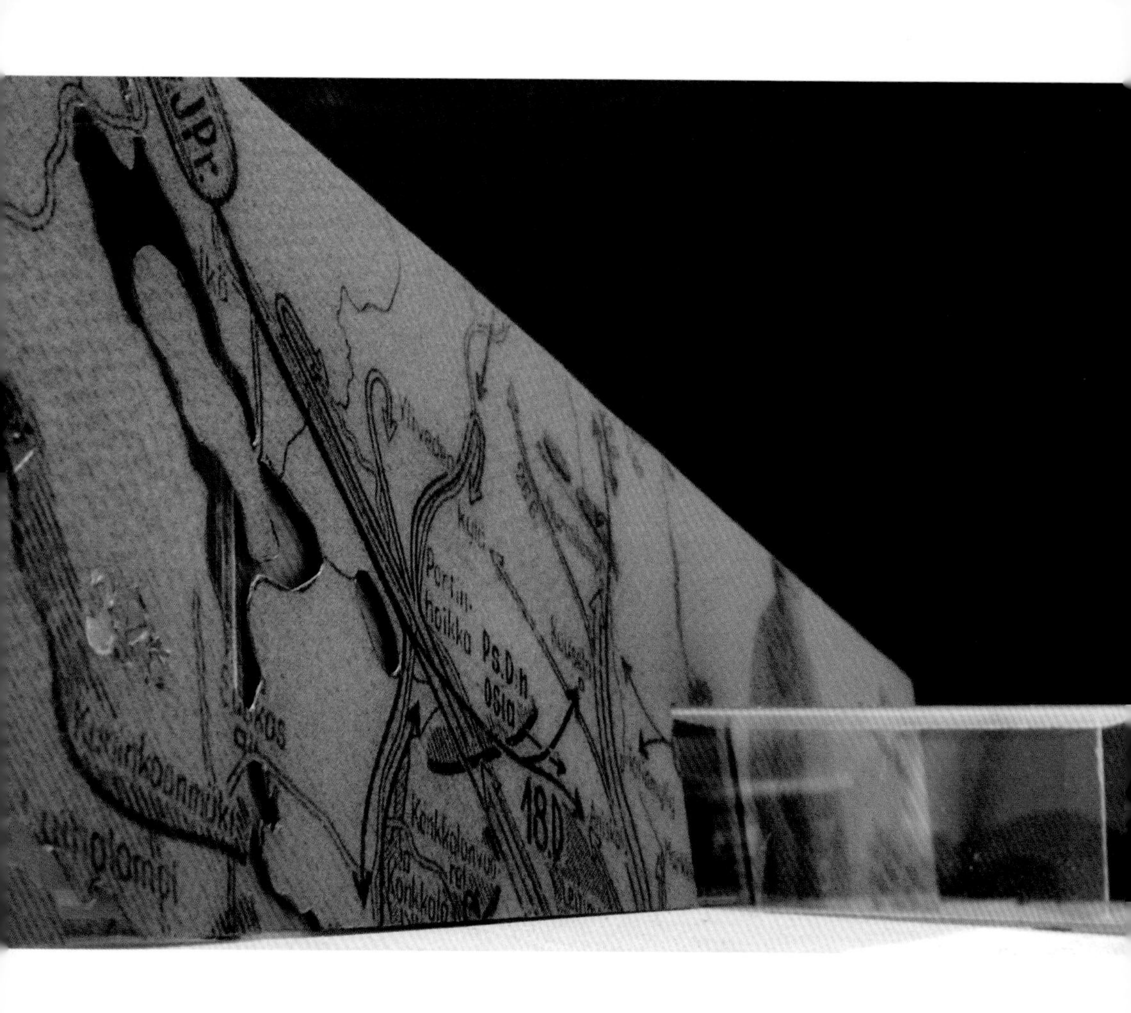
Portin-
hoikka
Ps.D:n
osia
18.D

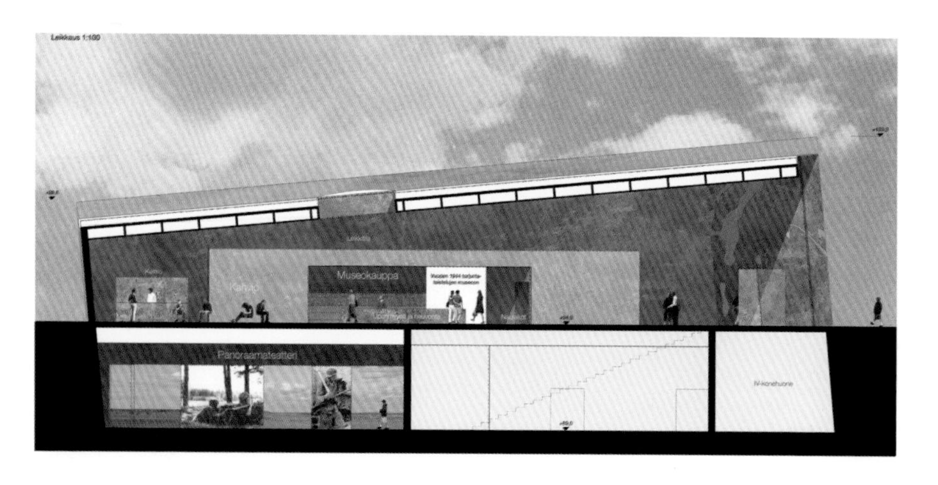
Leikkaus 1:100
Kahvio
Museokauppa
Panoraamateatteri
IV-konehuone

n the architectural competition proposal, the mall museum building presenting the histo-y of the decisive battle in the war between 'inland and the Soviet Union gives the im-ression of a bunker or a wreck partly sunk-n into the ground in the wake of a battle. 'he facades of the building are Corten steel, on to which are etched or cut the armies' groupings and novements. This was executed in the model by hotocopying the images on to brown cardboard nd cutting into it arrows representing the move-nents of the battalions. Transparent photocopies f similar tactical maps have been glued to the crylic sheets of the entrance facade.

反射和透明度

REFLECTIONS AND TRANSPARENCIES

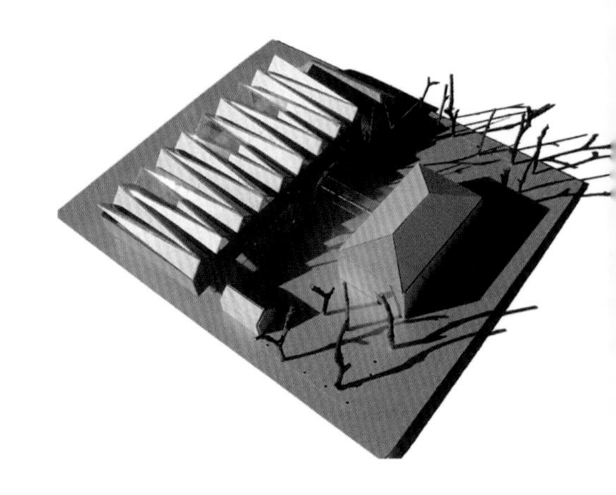

波尔沃市政厅

芬兰，波尔沃，2008年竞赛

Porvoo Town Hall

Porvoo, Finland, 2008 Competition entry

在老波尔沃市政厅的后面进行扩建，以作为该市的市政办公楼。方案中的屋顶形状受到了波尔沃河沿岸的河岸线仓库的启发。除了屋顶，该模型都是木质的，屋顶由金属与薄膜一起固定在折叠亚克力结构上，并置于坚固的底座之上。在不同光线条件下拍摄的模型照片，经由 Photoshop（图像处理软件）处理后用作效果图片，而不是使用 3D 建模软件制作的透视图。

An extension will be built behind the old Porvoo Town Hall for the city's civic offices. The form of the roof of our proposal was inspired by the shoreline storehouses along the Porvoonjoki River. The model is wood, except for the roof structure, where a metallic adhesive film has been fixed on to a folded acrylic construction that stands on the solid base part. The photos of the model, taken under different lighting conditions, were edited in Photoshop and used as illustrations instead of perspective drawings made with a 3D modelling program.

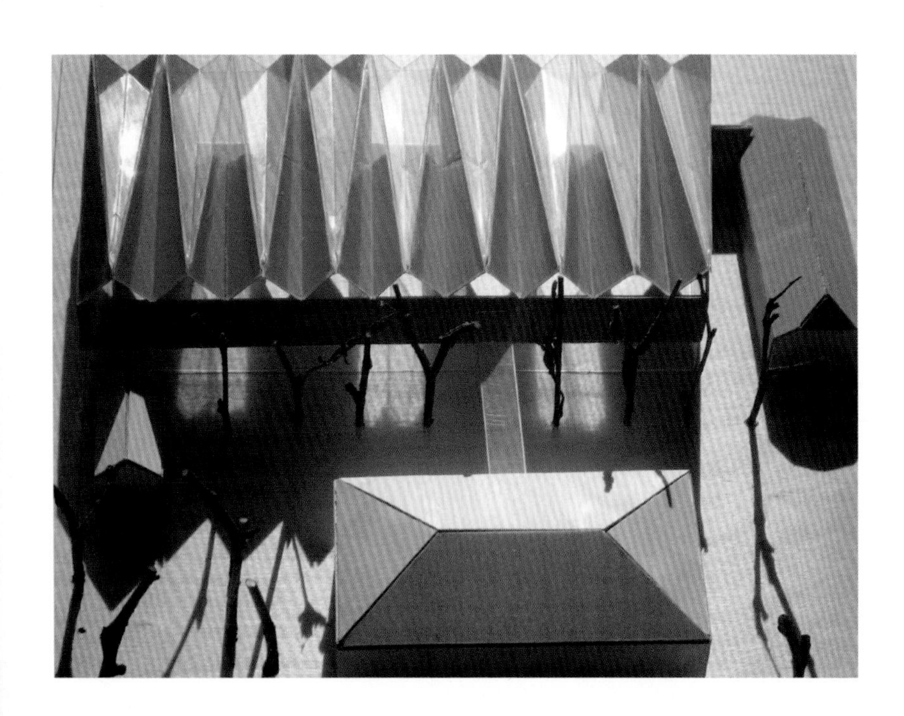

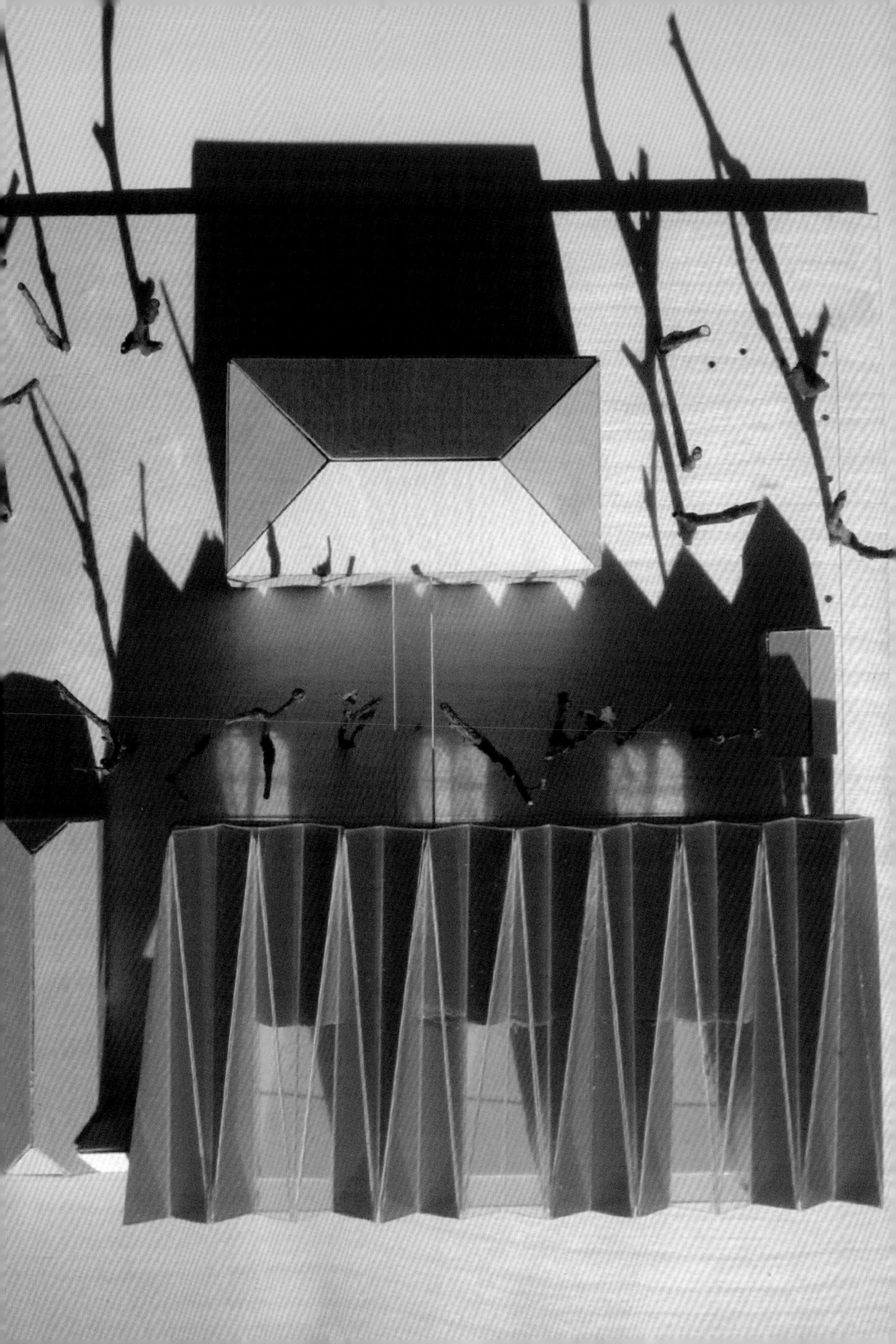

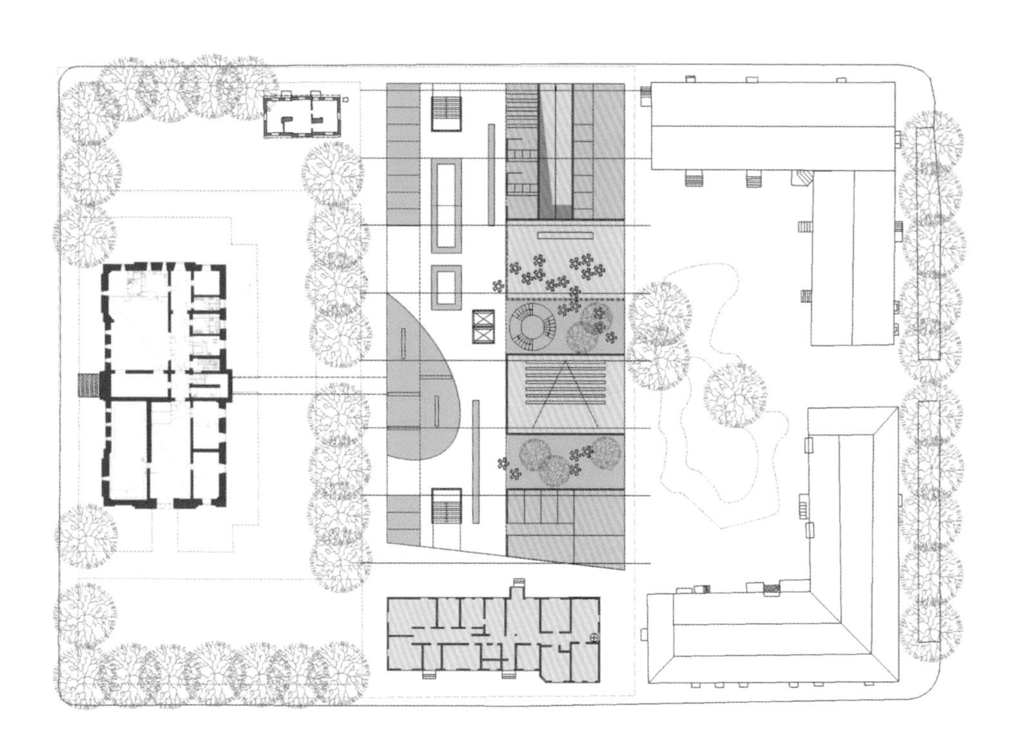

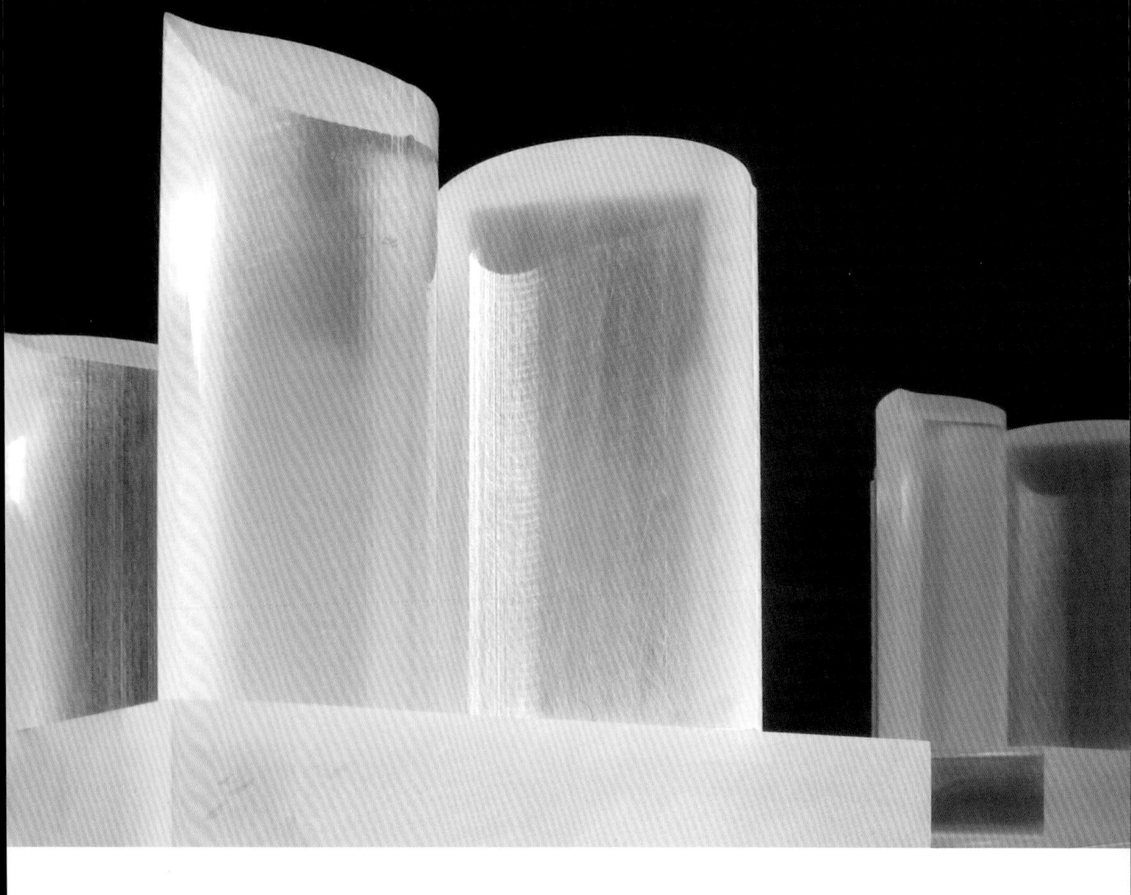

数控压磨亚克力

:NC-MILED ACRYLIC

每里恩公路项目

郡柏林，爱尔兰，2008年竞赛

Ierrion Road Project

)ublin, Ireland, 2008 Competition entry

这是一个位于都柏林市中心以外的城市街区建筑设计竞赛项目，建筑包括住宅和办公部分。方案的形式语言灵感来自爱尔兰国徽——三叶草。比例模型由固体亚克力制成，用数控机床加工成所需的形状。亚克力表面经过轻微打磨，在不同的光照条件下，材料的透明度和相邻塔楼的反射创造了一个千变万化的整体效果。

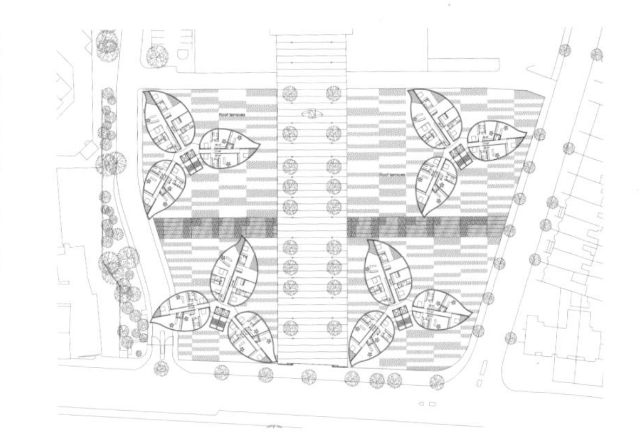

An architectural competition for the design of in urban block just outside the centre of Dublin :omprised of both housing and offices. The form anguage of our proposal took its inspiration from ι shamrock, the national emblem of Ireland. The scale model is made from a solid acrylic piece, which was shaped in the desired form with a CNC nill. The surface of the acrylic was lightly sanded so that the transparency of the material and the reflections off the adjacent towers create an ever-changing totality, depending on the lighting conditions.

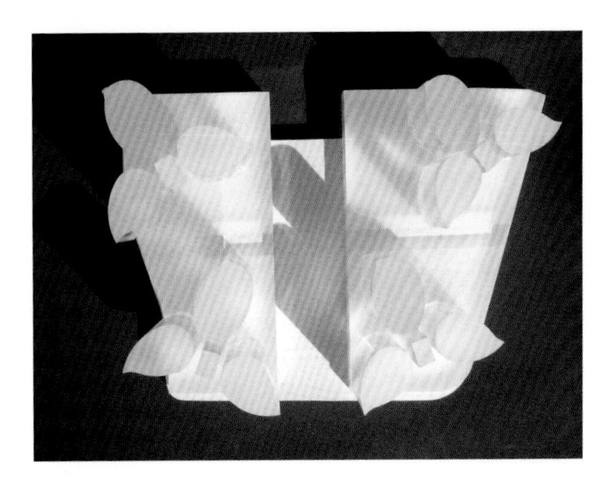

可看作一个真实的模型

MODEL BUILT AS A REAL ONE

坎纳斯塔罗拉瓦式活动板房

2007年

Prefab House Laavu for Kannustalo

2007

这个按照1：50比例建造的全木质模型，与它所代表的单户住宅的材料一致。这栋房屋的单坡屋顶由预制构件组装而成，其灵感来自传统的“拉瓦”[1]或由木料和泥炭建成的倾斜“棚户”。

This model in the scale 1∶50 is built completely in wood, as is also the single-family house it represents. The inspiration for the house's mono-pitched roof, assembled from prefabricated elements, came from a traditional “laavu” or lean-to shelter built from timber and peat. The structure of the model followed closely that of the actual house.

[1] 拉瓦是一种少数民族居住的传统单坡建筑，旨在为在野外徒步或钓鱼的人士提供临时居所。拉瓦常见于芬兰拉普兰地区，一般靠近受欢迎的钓鱼河流和国家公园旁。原则上，拉瓦是荒野小屋的简化版本。像荒野小屋一样，拉瓦不保暖；与荒野小屋不同，拉瓦没有门窗。典型的拉瓦是一个木制建筑，面积约10平方米，高2米，由屋顶、地板和三层墙组成，第四面墙永久是敞开的。——译者注

该模型的结构与实际房屋的结构很接近，大比例模型还可以显示内部空间的细节。屋顶是在亚克力板表面上切槽，而亚克力板下面的灰硬纸板则表明出屋顶的哪些部分是不透明的，哪些是透明的。该模型的屋顶是可拆卸的。

The large-sized model also allows for showing the interior spaces in detail. For the roof, grooves were cut in the surface of an acrylic sheet. The grey cardboard beneath the acrylic sheet indicates which parts of the roof are opaque and which are transparent. The roof of the model is detachable.

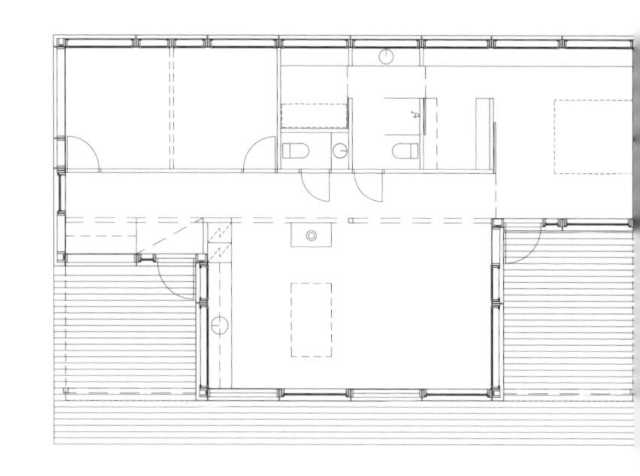

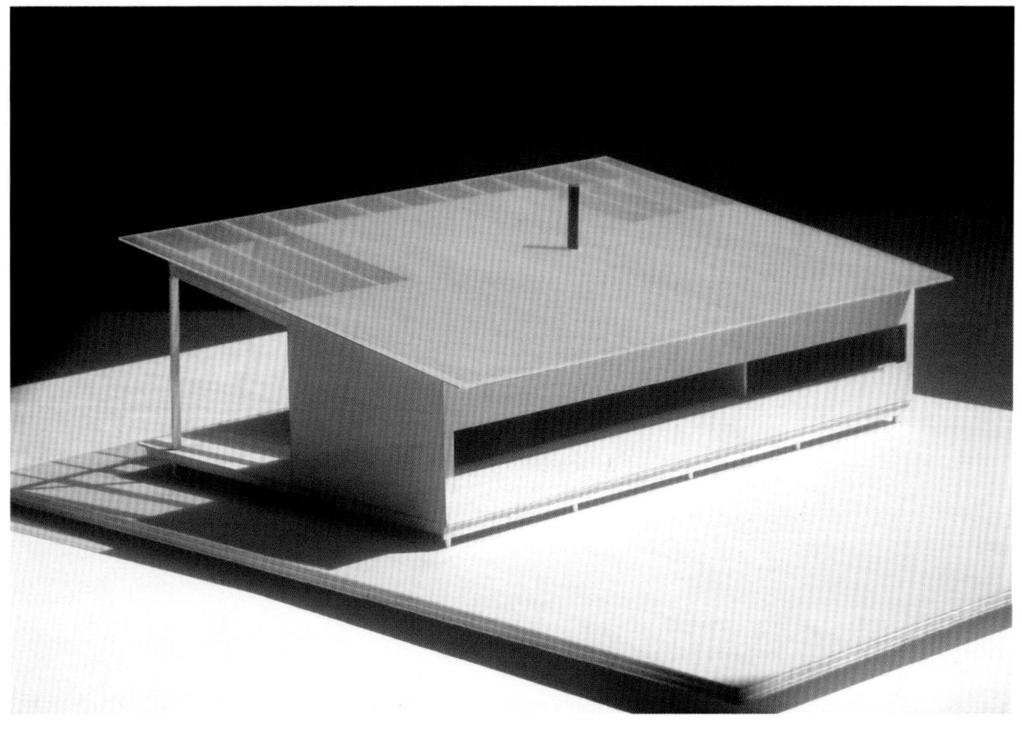

珀赫拉住宅区

芬兰，耶尔文佩，2007年竞赛

Perhelä Housing Block

Järvenpää, Finland, 2007 Competition entry

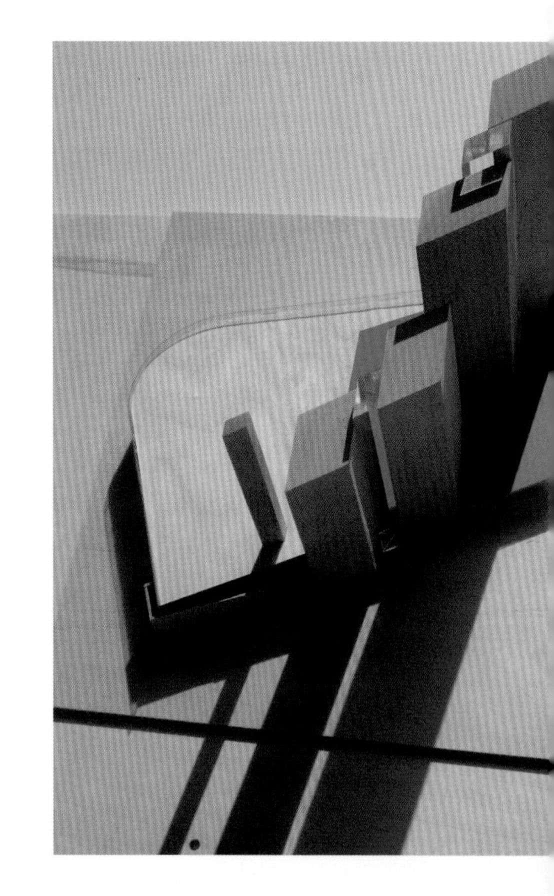

在此次建筑竞赛中，要把一个小镇的旧街区改造成一个当代住宅和商业中心。在方案中，两个住宅塔楼用玻璃楼梯间进一步细分成两个部分，从而形成六个垂直建筑体。这些塔楼被组合成为一个序列，向南逐渐增高，类似于墨西哥建筑师路易斯·巴拉干[1]（Louis Barragan）为墨西哥城设计的城市雕塑——卫星城塔（Torres de Satélite）。

In this architectural competition an old urban block in a small town was to be transformed into a contemporary residential and commercial centre. In the proposal, each of two housing towers was subdivided with glazed stairs into further two parts, which created a composition of six vertical elements. The towers have been grouped into a sequence that rises towards the south, in a way similar to Louis Barragan's urban sculpture Torres de Satélite in Mexico City.

[1] 路易斯·巴拉干（1902-1988）是墨西哥20世纪庭园景观设计的著名建筑师。巴拉干设计的景观、建筑、雕塑等作品都拥有一种富含诗意的精神品质。他主张将建筑与景观相融合，将建筑与景观作为一个整体进行设计。他作品中的美来自对生活的热爱与体验，来自童年时在墨西哥乡村接近自然的环境中成长的梦想，来自心灵深处对美的追求与向往。——译者注

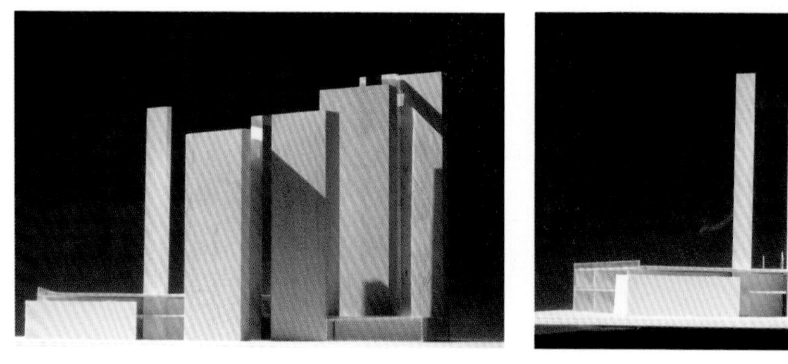

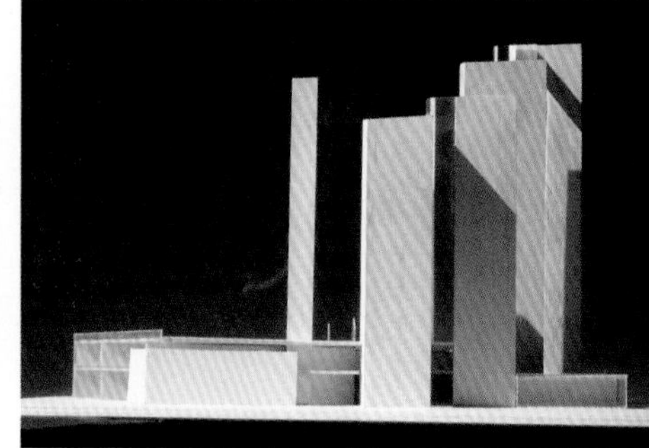

该模型由木块和亚克力块组装成所需的形状。塔楼建筑群给人的印象随着人们观看视角的不同而完全改变，这一点在比赛展板中通过一系列从不同视角拍摄的模型照片进行了充分说明。

The model was assembled from wooden and acrylic rods that were planed to the desired shape. The impression created by the tower composition completely changes depending on the viewing angle. This was illustrated in the competition display panels by a series of photos of the model taken from different directions.

与客户一起工作

WORKING WITH A CLIENT

内米斯托住宅楼

芬兰，赫梅琳娜，2007年

Niemistö House

Hämeenlinna, Finland, 2007

这栋单户的住宅是为一次住宅博览会而建的，其目的是展示不同的住宅布局。与靠近院子的一侧相比，临街立面更加封闭，仅通过一个私人露台向森林公园开放。

This single-family house was built for a housing fair, where the objective was to present different dwelling arrangements. The street facade is more enclosed compared to the yard side, which opens up towards a forested park via a private terrace.

该模型向客户展示了房间的布局概念，将自由形式的外部空间切入长方形建筑内，从而将更多光线引入住宅中，同时也便于眺望户外景观。建筑的承重结构为木柱和横梁，房间布局由白色硬纸板墙表示，被自由地放置于结构网格中。

The model showed the client the concept for the room layout, with a free-form exterior space cutting into the rectangular-shaped building, thus bringing light deep into the dwelling, as well as views outwards. The building's load-bearing structures are indicated by wooden pillars and beams, while the room layout, indicated in the model by the white cardboard walls, is placed freely within the structural grid.

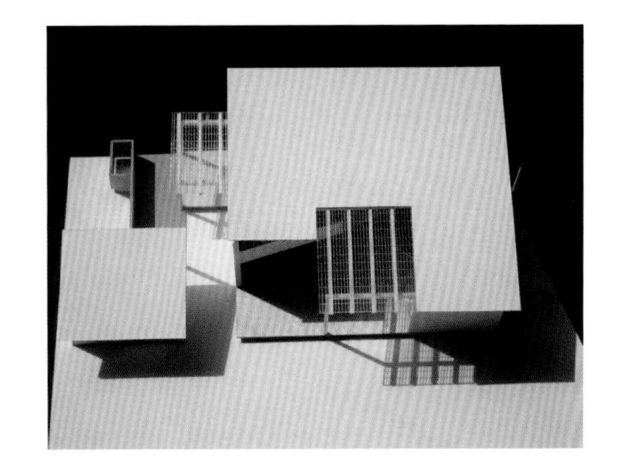

用模型做草图
SKETCHING WITH A MODEL

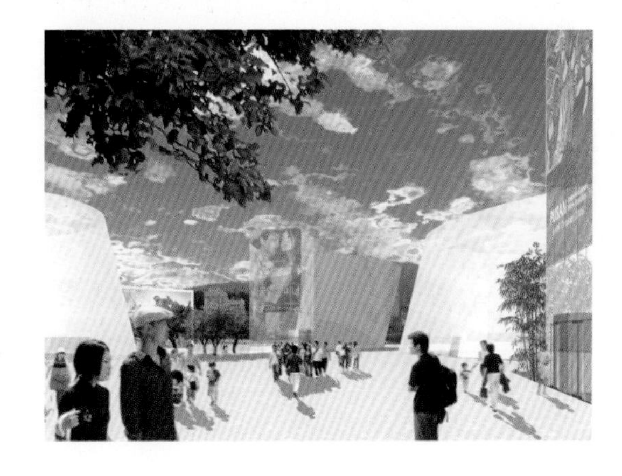

釜山电影院建筑综合体
韩国，釜山，2005年竞赛

Busan Cinema Complex
Busan, South Korea, 2005 Competition entry

釜山国际电影节主会场建筑综合体设计邀请赛于2005年举行。在我们的方案中，整个建筑分为两个层次：一个是光线充足的屋顶露台，其上放置了“电影城”及其剧院入口和户外屏幕；另一个是露台下面的电影剧院和国会大厦组成的建筑综合体。

An invited architecture competition for the main venue of the Busan International Film Festival was held in 2005. In our proposal the overall complex was divided into two levels: a light-filled roof terrace on which was placed the “film city”, with its theatre entrances and outdoor screens, while beneath it is a complex of cinema theatres and congress spaces.

借助于由白色硬纸板和亚克力制成的工作模型，探索了空间布局、外立面和建筑整体。模型从下方被照亮，因此通过模型照片可表现夜间的氛围。

The spatial layout, facades and the architectural totality were explored with the help of a working model built from white cardboard and acrylic. The model was lit from below, so that it was possible to visualize the night-time ambience through photographs of the model.

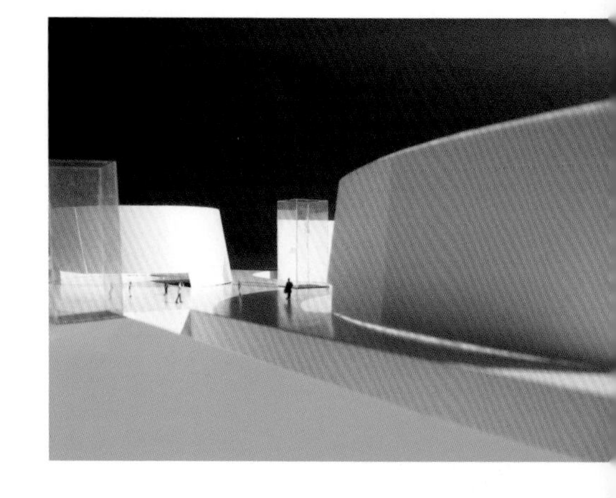

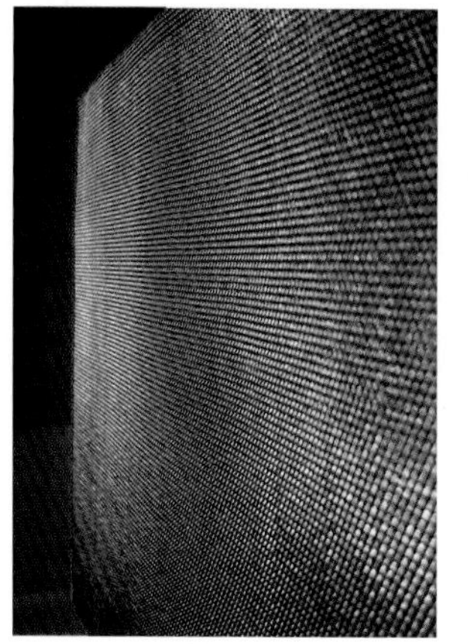

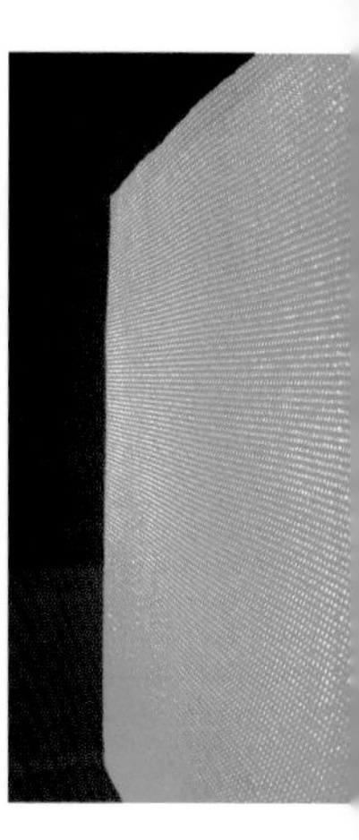

在模型中，基座的外立面由反光膜材料制成，其
前面是薄金属网。在比例模型中，我们想要表达
出立面创造的效果，根据从工作模型中获得的经
验，最终的模型委托其他模型制造商制作完成，
并提交参赛。

In the model the facades of the plinth are mad
from reflective film, in front of which is thin meta
mesh. In the scale model we wanted to show th
effects created by the facade. Based on the experi
ences gained from the working model, a final mod
el commissioned from an external model builde
was submitted to the competition.

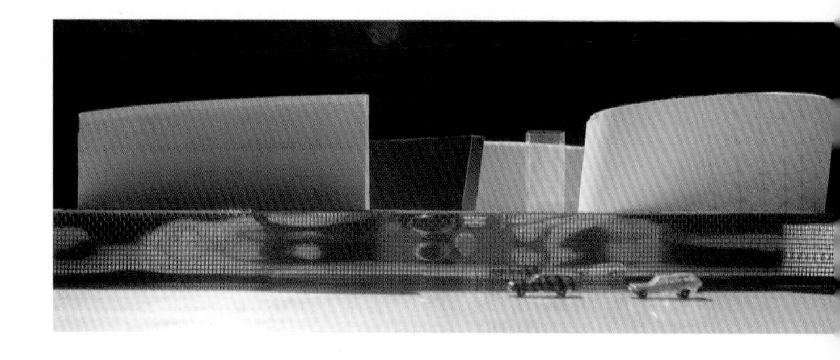

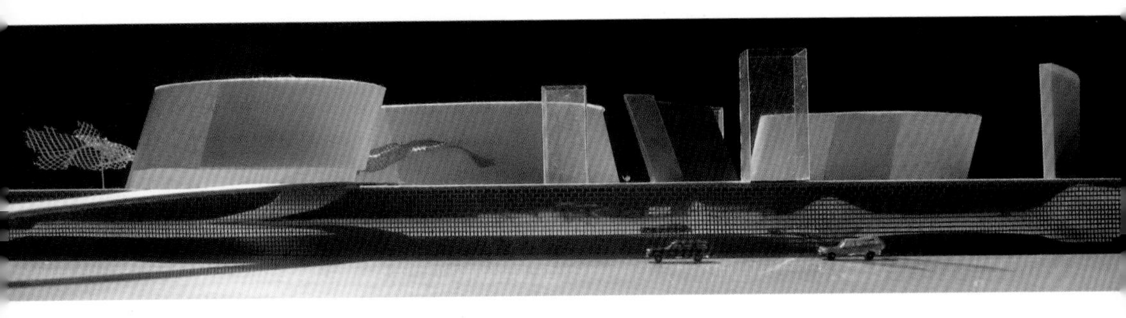

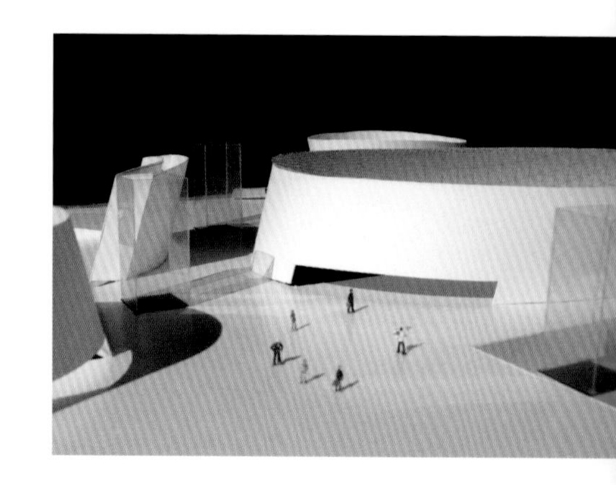

附录 1 | Appendix 1 序英译文

Model Thinking of Architects

From Abstraction & Geometry to Architecture Masterpieces

Fang Hai

Heikkinen-Komonen Architects is the most active and representative Finnish architect group in the contemporary world. Its two founders and also main partners, Mikko Heikkinen and Markku Komonen, have served as architecture professors at the Architecture School of Aalto University (the original Helsinki University of Technology) for long time. They are the typical representatives of scholar architects in Finland, and have represented a successful element of architectural education of Finland. The majority of employees are their former students, many of whom separate out to set up new design studios and larger architecture offices, which creates a powerful impression of endless inheritance and architecture vitality in Finland. Simultaneously, during the winter and summer vacations, college teachers as well as students in the fields of architecture or design from all over the world visit their office and exchange information with them. In the summer of 2018, the author led a group of architecture teachers and students from China to visit Heikkinen-Komonen Architects and got a warm reception from Prof.Heikkinen. Students were attracted by ubiquitous building models in the office, which were made of a variety of different materials with different proportions and styles. The models were full of extraordinary creativity and attraction, just as a small architectural museum. Students, being surprised and joyful, saw not only a lot of building models made of different materials in various design stages but also many kinds of manufactured models and essential tools on the design tables. Then, the following conversation started.

Students(S): Prof.Heikkinen, is making models your daily work?

Prof.Heikkinen(H): For me sketching with models is essential at the early stage of the project. Shaping simple compositions using materials to be found on your tabl like tracing paper, cardboard and transparency film is part of my thinking process. Even a most primitive piec of folded paper can convey the basic idea of a possible solu tion. With cardboard and glue, it is sometimes faster t demonstrate your design than through 3D screening.

S: So, are all the models in your office made by you an your colleagues?

No, most models you can see here at our office are mad by professional model makers. These models are mad after the actual design process is already closed or eve when the building or project is finally implemented. The are built for a client review or as a part of a competitio entry or for an exhibition. We do not have the expertis or machinery to make that kind of perfect objects. Man of the sketchy models made by us in the beginning of th design ends after our review and reject process into tras bin under the worktable.

S: It seems that hundreds of models in your office are mad of all the materials we can imagine. Do you have require ments for models materials? For instance, how do you as students to make models in your teaching process? Are ther some limitations and recommendations to the materials?

Any material at hand is ok. The effort to sketch 3-dimension ally with your hands should be as easy and fast as possible But the more elaborated your idea gets, the more articu lated your model could be. Using different materials, yo can characterize your design better. In the beginning i may be enough to study the form of your object, then yo may, for example, emphasize transparent and solid part of the volumes. Details are not important in this phase you should strive to strengthen your overall approach.

Ve try to push our students to start playing with models efore they dive into digital world where they usually get asily lost. I think it is healthy not only to stare the screen ut to work also with your own hands.

NOTE: the last Q & A is deleted!

. Bad Experiences of a Finnish Architect's Abstraction & Geometry n China

The maximum number of buildings in Chinese urban contruction appeared in the past 30 years, and architects rom all over the world swarmed to China to participate in lesign and building, who, together with large number of Chinese architects, have greatly improved the appearances of a lot of Chinese cities. And then they promoted Chinese urbanization with incredible construction speed. In general, architects, from Europe, America and Japan, coninuously introduced the most fashionable architecture deation and the most advanced architecture technology and materials for Chinese architecture, and simultaneousy, they promoted Chinese architects to grow together during the cooperation process. However, until today, when ve recall and reckon the most outstanding and striking buildings in Chinese cities, most of them? Are designed by architects from Europe, America, Japan, Hong Kong and Taiwan. What is the main reason for absence of Chinese architects? There are some reasons, relating to management system, history and education, but the primary one s that Chinese architects lack design philosophy and concrete design techniques.

Pekka Salminen, a Finnish architect master, was one of the first Nordic architects who entered China and one of the western architects who succeeded in China. In the past two decades, Salminen Architects participated in more than 100 international architecture competitions in China and won the first prize more than 20 times. It completed a large number of construction projects, including Wuxi Grand Theatre, Chengdu Yunduan Building, and Fuzhou Strait Culture and Art Centre. Those large public buildings have become architecture bench marks, and have been highly rated by owners and architects from home and abroad. Prof.Salminen, 80 years old, had a lot of emotions whenever we talked Chinese contemporary architecture and his design practices in China during the past two years. He talked about the profoundness of Chinese traditional culture and also commented on enigma of Chinese officials. He often admired the amazing design talents of ancient China, and also sighed with nearly devastating impacts on culture and traditions from quick success of contemporary China. For Chinese contemporary architecture and architects, Prof. Salminen sang high praises for their powerful learning abilities, but meanwhile, he also kindly pointed out the deviation of their design philosophy, especially the lack of abstract ability. Actually, insufficient status of abstraction does not only exist in Chinese architects but also Chinese officials and ordinary people who are generally accustomed to concrete thinking, such as urban sculptures throughout China, and thus vividness is highly accepted. Prof.Salminen acknowledges that insufficiency of abstract thinking ability limits basically the potential intelligence of Chinese architects. Because the Chinese architects don't put emphasis on geometry, scale and proportion, they make monotonous designs so as to decline of the quality of general buildings.

Prof. Salminen admits that maybe because Chinese concrete culture with a long history makes a deep impression in public hearts, Chinese architects and city officials believe that symbolized ping-pong concrete thinking always easily overrides the abstract one. Prof.Salminen loves Chinese culture and often incorporates many design elements with Chinese culture symbols. These Chinese unique design elements are filtered into a series of new creative symbols in Finnish architects' minds, which are integrated with long-standing geometric design techniques in western cultural traditions. Then, Salminen's works, such as Wuhan Hubin Hotel with Wild Geese Figures Alighting on Sand, Wuhan International Airport with Hawk Flying Wings, Wuxi Grand Theatre with Dragonfly Gracefulness Walking on Water, and Fuzhou Strait Culture and Art Centre with Sailing Boat Race, immediately won owners' praises, and some of which were masterpieces with Finnish symbolic design and eventually were evolved into great landmarks in Chinese cities. However, obvious symbols for modern urban architecture are not always the best choice and in most cases even an improper choice because high development of human science and technology originate from rational abstract thinking, so modern architectures, closely related to high technology and new materials, must be dominated by abstract thinking. The leading role of abstract thinking is not only reflected on types of buildings, such as skyscrapers and large-span buildings, supported by high technology and high-strength materials, but also on small-and-medium-sized culture buildings and landscape architecture. Prof.Salminen repeated to author about his terrible experience of Finnish Architect's *abstraction & geometry* in China, which included his confusion and helplessness. Prof. Salminen once explained roof structure concept of local civil exhibition hall to a deputy mayor by means of quick sketches. As a result, the executive official, listening carefully, unexpectedly asked questions about how to deal with those broken lines in sketches during actual construction process. Although Salminen explained that they were just imaginary sketch lines and all broken lines would be filled up in later stages, he still sighed that Chinese city officials lack abstract thinking. Once again, Salminen Architects participated in an architecture competition for some community school of Shanghai and entered the final stage. Prof.Salminen's scheme fully demonstrated the top level of architecture design in Finnish contemporary school with delicate geometric composition throughout the whole layout, elevation and spatial arrangement, which has been confirmed as an implementation by review experts. However, at the final demonstration meeting, it happened that a supervisor who was interested in this scheme suggested to add a lot of symbol elements in plane and facade at the expense of reasonable function configuration and smooth spatial organization. This leader's suggestion eventually resulted in project failure and was accepted by owners who gave up an excellent scheme but would rather chose a vulgar one, full of symbols but without functions. Prof.Salminen was repeatedly helpless. Many European and American architects the author knows encountered the similar experience in China. This is the only way for Chinese modern ultra-construction process which will bring out inevitable contradictions and difficulties.

It is gratifying that rapid development of network recently enabled architecture criticism directly to enter huge numbers of families. Various civil organizations often not only report about hundreds of constructions like little White House as well as commercial streets imitating Ming and Qing Dynasty in China, but also select the top ten ugliest Chinese buildings every year, from which we can find Chinese transformation and improvement of aesthetic taste. When the public mock buildings such as Fulushou Hotel, Golden Yuanbao Building, Huge Copper Money Arch, Big Teapot, and Large Iron Pot in some places, we have clearly found that symbol images are turning to abstract thinking and majority of Chinese architects also increasingly focus on training of abstract thinking through geometry, and gradually enter the greatest architecture stages with scale and dimension as core space soul.

2. Two Books, Origins of *Classical Architecture* and Theory on Architecture

A new book, *Origins of Classical Architecture: Temples, Orders and Gifts to the Gods in Ancient Greece*, written by Mark Wilson Jones, was published by Yale University Press, which, from architect's perspective, attempts to explore the most representative contents of continuing classical architecture tradition in the western world: classical orders namely Dorian, Lonic and Corinthian. The ancient Greek architecture was built on the basis of ancient Egyptian architecture, followed by ancient Roman architecture, and then it started the Italian Renaissance together with ancient Roman architecture which afterward led to the European Renaissance and finally led global civilization development. The core elements of classical architecture from ancient Egypt to ancient western Asia to ancient Greece and ancient Roman were classica columns which were the soul of structure system in a classical architectures. Their forms directly derived fror nature, especially from woods, namely abstract and sim ple tree forms. Those abstracted trees were geometricall planned by architects through long-time evolution in an cient Egypt and ancient Western Asia, which finally be came the western classical column systems at the age c ancient Greece, represented by the large existing numbe of instances, such as Dorian, Ionic and Corinthian. Then they were all accepted into ancient Roman architectur systems and became core elements of classical architec ture systems in Europe and America. Thus, the wester architecture tradition was firmly established on the bas of *abstraction & geometry* thinking pattern. The ration al thinking coincided together with industrialization an modernization of mechanical era to drive and promote th modern architecture triumph.

Professor Jones, after long-term investigation and re search, found that maturity and form of ancient Greek col umns were completed in a relatively short period and wer closely related to the development of local culture, inter national communication, and artistic innovation desire Most of ancient Greek buildings which were passed dow after generations until today were temples, undoubted ly the most important buildings in Greek society at tha time. Ancient Greece was a society with humans and Go living together, thus, lofty thearchy formed in this way no only possessed high position but simultaneously was close ly related to the human society. Therefore, ancient Greel temples achieved a unique-and-sacred status. Temple wa a place where all human beings consecrated themselves t thearchy, and its building itself was also the most signifi cant to thearchy. Therefore, tall height, solemn columns and luxury structures were necessarily accepted to forn basic temple appearance demonstrating luxury, loftiness and long-time life and to create a new architecture forn in the western classical world. Classical columns with loft iness and sense of religion and temple forms eventually became continuous venation soul of the western architec ture, which remained unshakably for thousands of year in the Middle Ages and officially became European archi tecture heritage through Renaissance. Its core idea wa *abstraction & geometry*, which was completely compatible with development requirement of modern metropolis afte industrial revolution, and thus became core idea of mod ern architecture and design. In contrast, wooden structure systems in China invented by our ancestors were based or column structures, equally exquisite and luxurious witl stunning to the world in the past dynasties. However, th weakness of the structure was easy to be burned and diffi cult to last. Brackets, the barely corresponding structure to the western classical columns in ancient Chinese archi ecture, connect structure elements as well as intellectual ames with tactics and complexity. With time going, its ecorative function gradually exceeded the meaning of ructures. Some scholars believed that the emergence f brackets derived from imitation of tree canopy, but it icked *abstraction & geometry* thinking basis of the western classical columns, which was difficult to be integrated into modern society and also the weakness of Chinese modern architects too. Beginning with Liu Dunzhen, iang Sicheng, Yang Ting bao, and Tong Jun, generations f Chinese architects have been exploring successively, but most of them were experts among national and international styles, urbanization and localization styles, but few f them really paid attention to *abstraction & geometry* of rchitecture basic thinking pattern, which always existed n Models made at Heikkinen-Komonen office are really resented among excellent buildings of Finland, Europe, merica, Japan and other countries, which is worthy of eing thought repeatedly by Chinese architects.

heory on Architecture: Form Renaissance to Today, written by Bernd Iverson, a German scholar, was translated by ang Yun, and was published by Beijing Publishing Group ompany in June, 2018. The first edition was published n 2003 by the famous Taschen Press with more than 800 ages and a large number of original illustrations introducing architecture and related discussions from elites f various countries in detail, such as Italy, UK, France, ermany, Spain and other countries. In the west, the roots f architecture theory were generally traced back to *Ten Books on Architecture* written by Vitruvius, an architect nd engineer of ancient Roman, which was the oldest architecture book and was reprinted in various languages. itruvian's works recorded the main architecture types f the ancient Roman era, thus faithfully conveying abstract thinking tradition and geometric representation f the western architecture with classical columns as its ore. *Ten Books on Architecture* was not only precious inrmation about the western classical architecture and its rinciples but also was the cornerstone of all architecture inking and theories in the western world especially after enaissance.

n the introduction part of the book, Christopher Tools tarted topic from Architecture and Literature and introduced Humanism, Renaissance, Palace Culture and City ulture, Books, Illustrations in Books, Column Types, Architecture Theory and Utopian Society, Theory and Couner-Theory and so on. In terms of the western architecture evelopment itself, column types were the most significant barometer. On the one hand, it represented cornertone elements for the western architecture. On the other and, it also stood for the development and evolution of rchitecture in different regions and countries during estern-world in various eras. No matter any alteration, *abstraction & geometry* thinking pattern symbolized by column types never changed. Architecture development in European countries was basically synchronized before modern society based on development process of architecture theory. From Alberti, Vignola, Palladio to Piranesi in Italy, their architecture theories mainly focused on classical architecture pattern, particularly the revival and standardization of column types. Architects such as Honnercourt, Perot, Blondel, and Le Duke in France, specifically applied and developed column types by more practical instances. In the United States of America, Shute, Chambers, Putin, and Ruskin stressed on surveying and mapping ancient architecture sites and recreating ancient columns through recording. German architects, such as Durer, Grunte, Shenker, and Semper, not only exclusively recorded the most important architecture examples of ancient world through precision of German style and the delicacy of copperplate, but also deeply discussed techniques of various perspective presentation. On the one hand, the architecture theories in the 20th century were established on the basis of the western classical traditions. On the other hand, they were related closely to the development of new technologies and new materials in new era. Architects, from Howard, Loos, Tauth to Le Corbusier, Walter Gropius, Wright, then to Michael Dion, Wenjeri and Koolhaas, enthusiastically discussed the popular architecture topics and attempted to provide some suggestions and answers. However, their discussions and explorations never took apart from rational thinking laws of *abstraction & geometry*. The topic of *Theory on Architecture* was to continually discuss Greek column styles, and to keep the rational thinking pattern during discussion which took *abstraction & geometry* as its core element. This was the most significant cornerstone of modern architecture, which should arouse a new round of discussion about Chinese traditional architecture heritage among Chinese architects.

3. Creative Soul of Modern Architecture Master: *Abstraction & Geometry*

Every master in modern architecture movement developed and enriched modern architecture through their own different ideas and techniques with their common creative soul: *abstraction & geometry*, which were both ideas and techniques and thus formed the main melody of modern architecture development. In a sense, each generation of modern architects interpreted the meaning of abstraction with their own unique way and simultaneously transformed modern technology and materials with geometric techniques into space beyond their times.

At the age of 75, Wright finished his autobiography which was continuously republished after the first version in1932 until the final one in 1943, containing all five volumes. The autobiography was subsequently published in various languages all over the world. Shanghai People's Publishing House also published *An Autobiography: Frank Lloyd Wright* translated by Yang Peng. It included five volumes: family, connection, career, freedom and form. Wright himself designed abstract line patterns for each illustration in every volume, revealing his skilled mastery in presenting geometric patterns. Wright, in all his life, no matter in town planning or architecture, in interiors or homes, in carpets design or household utensil design, consistently comprehended and created every aspect of human habitation and their surroundings based on his understanding of geometry. In 2005, famous British press, Phaidon, published a new book named *On and By Frank Llond Wright: A Primer of Architectural Principles*, chiefly edited by Robert McCarter, an architecture professor of Washington University, which included three essays written by Wright himself: *In the Cause of Architecture, The Logic of the Plan,* and *In the Nature of Materials.* These essays directly expressed the design philosophy of Wright himself: the inspirations and enlightening elements were drawn from nature, and then were presented as space logic by geometry and abstract thinking patterns. When appropriate and ecological materials were applied to present space with logic relationship, a masterpiece of modern architecture coexisting harmoniously with nature would be finally created. Other essays in this book from fourteen scholars such as Kenneth Framption and Colin Rowe, were about their narrations on hundreds of sketches and attached drawings from archives of Wright, which clearly stated how Wright thoroughly studied geometry theory and applied them to the evolution process of creating spaces.

Bauhaus-Universitaet Weimar, the birthplace and base of modern architecture and international style, is also a perfect example of thorough advocating and practicing thinking pattern of *abstraction & geometry* in the field of modern architecture, design and art. The three original presidents of Bauhaus were all world-class architecture masters, all of whom made landmark achievements in city planning, architecture and design. Walter Groplus, the pioneer and the first president of Bauhaus, had exceedingly profound understanding of relationship between abstract ideas and fields of architecture, design and art, whose architecture works were remarkably successful. As a result, he instantly became spiritual representative of new era. The second president, Hannes Meier, devoted himself to integrating *abstraction & geometry* thinking patterns to social reform and collective housing construction in urban and rural areas. And the third president, Mies, took the lead to apply abstract collage concept to presenting his advanced conception of flowing space and glass skyscrapers. Meanwhile, he set up architecture details by extremely strict geometry techniques to create construction stan[illegible]ards for modern metropolitan cities. The fourth worl[illegible] class master architecture master, Marcel Breuer, showe[illegible] his foresighted innovation consciousness in two fields [illegible] construction and furniture. He was a superb practitio[illegible]er in applying abstraction-and-geometry model, and als[illegible] created a new shape language of modern architectur[illegible] The group of Bauhaus masters provided epoch-markin[illegible] understanding and application of *abstraction & geomet[illegible]* thinking patterns, which were also a natural outcome du[illegible] to Bauhaus as a birthplace and important place of mo[illegible]ern abstract art, where the most important pioneers an[illegible] top masters, such as Mondrian, Malevich, Van Doesbur[illegible] Lissitzky and so on, once lectured there. Other importa[illegible] pioneers of abstract art, such as Kandinsky, Klee, Moh[illegible]ly-Nagy, Albers, Feininger and Schlemmer, were full-tim[illegible] lecturers at Bauhaus. They were all inspired by Gropi[illegible] to lecture at Bauhaus to find abstract mysteries of natu[illegible] with different perspectives and then to apply geometr[illegible] color, materials, and images to construct buildings of a[illegible]stract art and simultaneously to affect artistic creatio[illegible] and design thinking from all over the world. Althoug[illegible] Bauhaus only existed for nineteen years, it was etern[illegible] fire of modern art and architecture. Masters and studen[illegible] of Bauhaus established Ulm School of Design, Monten[illegible]gro Art Academy, and Chicago New Bauhaus in Euro[illegible] and the United States, thereby to spread abstract thinkin[illegible] pattern and design creative ideas across the world.

After initiating a new area of modern architecture by fou[illegible] architects of Bauhaus, Le Corbusier worked as both an ar[illegible]ist and an architect. Artistic originality, as an internal wa[illegible] guided his work as an architect. The purism established b[illegible] Ozanfant and him together was enlightened by the Cu[illegible]ism, but was integrated with three-dimensional space[illegible] which was actually a series of object-image representatio[illegible] developed by *abstraction & geometry* thinking guidanc[illegible] Le Corbusier's early architecture was represented by Vi[illegible]la Savoye, as a spatial organization and facade evolutio[illegible] guided by abstract Europe-style geometry. His post-arch[illegible]tecture career was represented by Chapelle de Roncham[illegible] which was a multiple trans-boundary space constructio[illegible] and self-created façade evolution. Central Complex of Ne[illegible] Chandigarh, India, representative work of his later arch[illegible]tecture career, was a perfect combination of abstract co[illegible]ception and geometric dimensions from Le Corbusier[illegible] Corbusier's modulus-system study throughout his who[illegible] life was his theoretical induction and scientific summa[illegible] of *abstraction & geometry* thinking pattern for mode[illegible] architecture. His fine modulus study had multi-level an[illegible] multi-direction comparative analysis between archite[illegible]ture space, proportion, scale and humans, which helpe[illegible] his architecture works be filled with creativity and enligh[illegible]enment, durability, thoughtful consideration and enti[illegible] timelessness. Although Le Corbusier, one of the modern painting masters, occupied a place in the field of modern paintings and sculptures, he engaged himself permanently and deeply in modulus study, taking on a scientist-working attitude with expectation to reach scientific level. He had specially gone to Princeton University to visit contemporary scientific master, Einstein, with his manuscript of *Modulus*, and was encouraged and praised by him. Le Corbusier started his scientific research on architecture, city, daily design, and *abstraction & geometry* thinking pattern in 1912, the year to publish his book named *A Study of the Decorative Art Movement in Germany*. His collected papers, *Towards a New Architecture*, published in 1923, established his position of a chief spokesman of modern architecture. His systematic science research of architecture lasted for all his whole life, and he continuously published several books, including *One House, One Palace: A Study of Architectural Integrity, Yearbook of Modern Architecture, Art Deco Today, Planning of Three Human Settlements, The City of Tomorrow and its Planning, Accuracy: State of Construction and Urban Planning Reports* and so on. Le Corbusier's investigative architecture practices set a perfect example for later architects. Along with modern architecture stepping into stages of diversification, globalization and informatization, architects' understanding and research attitudes towards architecture from all over the world were also multiplex. On the one hand, architects after Le Corbusier and Bauhaus found shining examples, and on the other hand, they started to experience and explore creative approaches which suited themselves.

Louis Kahn: In the Realm of Architecture, written by David B. Browning & David G. Delon and translated by Ma Qin, was published by Jiangsu Phoenix Science Press in 2017, which showed us another approach of modern architecture. Winterson Skoley, a famous American architecture historian, precisely illustrated how Louis Kahn found his own architecture topics in his perennial teachings of architecture history in the book introduction. "Kahn found how to transform Roman ruins into modern architecture in his later life, and this relationship seemed apparently impossible, but he successfully achieved this transformation in all the architecture works after leaving from the Institute of Salk Biology." During his period of architecture-history teaching, Kahn repeatedly visited ancient architecture sites in Egypt, Greece, and Italy, experiencing soul shocks form classical and timeless *abstraction & geometry* sequences in ancient buildings. In his subsequent architecture practices, he directly introduced Egyptian pyramids, ancient Greek temples, and column spaces of Roman forum into his modern architecture projects. "Above all, Louis Kahn's 'sequence of bricks' applied in Ahmedabad of India (Indian Institute of Management) and Dhaka of Bangladesh (National Parliament House) came from Roman bricks and concrete structures, which were extracted from Piranesh's creation, while the porch of the Dhaka clinic resembled Le Duk's painting with architects' full view."

Ieoh Ming Pei was taught by Walter Gropius and Marcel Breuer in his earlier years, and learned the essence of modern architecture concept and had a very clear understanding of *abstraction & geometry*. Traces of "geometric abstract concrete" modeling of Breuer could be found in his earlier works. East Building of the National Gallery of Art in Washington, his first masterpiece, developed Breuer's design legacy to its full and simultaneously began to establish his own concept about geometric composition and abstract spatial scale. Later, he continuously added modern understanding of historically classical architecture sites in his following design practices, forming a kind of design style with strongly historic and cultural atmosphere. This world-shocking design style was first unveiled at Louvre Museum in Paris, whose reconstruction attracted worldwide attention. Ieoh Ming Pei knew that ordinary design theme and method could not be convincing, so he presented the most famous ancient Egyptian Pyramid in the architecture history. Unlike Kahn's direct space borrowing in classical sites of ancient Egyptian, ancient Greek and ancient Roman, Ieoh Ming Pei borrowed geometric proportion of Pyramids, and then chose the most popular materials of new age, mainly steel and glass, to make series of pyramids in Louvre square. This scheme immediately caused a worldwide sensation, and French attitudes changed from hate, uncomprehending, to love, which, from another point of view, reflected Ieoh Ming Pei's audacity and excellence in his aspect of creating techniques. As a Chinese American architect, he was invited to design the Fragrant Hill Hotel in Beijing at initial stages for reform and opening-up, who logically became the first international architect entering Chinese architecture design market. During the design process of Fragrant Hill Hotel, Ieoh Ming Pei transformed his creative thoughts to Chinese traditions of architecture and garden, but he interpreted them by *abstraction & geometry* thinking pattern of modernism. Therefore, we could enjoy a modern Fragrant Hill Hotel with distinctive Chinese garden style. In his later years, Ieoh Ming Pei, a representative of international architects of a whole generation, achieved the highest level of design skills. His skilled abstract concepts and geometric space techniques digested with different cultural traditions around the world, and produced a series of classic cases of abstraction-and-geometry modern architecture with Pei's style, such as Miho Museum in Japan, Suzhou Museum in China, and Museum of Islamic Art in Doha of Qatar and so on.

Contemporary architects realized deep implication of abstract concept and geometric dimensions from the creative ideas and design practices of senior masters, and then based on which they started to explore their own cultural characteristics and professional background to create architecture styles of their own. Botta, a Swiss architecture master, was edified by ruins of ancient Roman and ancient Greece in his childhood. Thus, he obtained inspiration from ancient Greek column styles and ancient Roman arches, and then made a modern geometric arrangement with bricks to serve modernization as Kahn. Tadao Ando, a Japanese self-thought master, was deeply infatuated with Le Corbusier's modulus study during his long-term investigation of European architecture, and was even fascinated by the concrete masterpiece of Le Corbusier and other Nordic famous architects. Therefore, he made a firm and insistent decision: He would build modern concrete buildings with *abstraction & geometry* space-shaping concept, and he was determined to study and create concrete materials to perfection with *abstraction & geometry* thinking patterns. Zaha Hadid, the deceased outstanding contemporary female architect, started from mathematics to modern architecture palace. Hadid became a young mathematician with much success in college, but she had a passion for architecture and painting. Hence, she entered the Architectural Association School of Architecture to learn architecture. She developed a comprehensive construction drawing collection starting from numerous laws and concepts of mathematics only understood by very few people, in which Non-Euclidean geometric elements forecast a series of architecture masterpieces would be born after a short while. By the means of modern computers and engineering development, she, in the query and wonder from all over the world, let herself integrate into her building schemes with Non-Euclidean geometry. She succeeded in building the Vitra Fire Station with concrete in Germany, and since then her amazing architecture talent could not stop any more. Assisted thoroughly by steel and glass, Hadid was really excelled in architecture realm introduced by mathematics. Daniel Liebskind, another extraordinary architecture genius, entered a unique drawing collection beginning from different points, and then imported architecture practices. Liebskind read many books, particularly in architecture and art history, and was employed by Cranbrook Academy of Art established by Eliel Saarinen. In addition to the relaxed and pleasant teaching, Liebskind started to develop a set of three-dimensional geometric attractions based on Islamic detailed points and architecture points of general renaissance, which, so unique and distinct as a drawing collection, was very fascinating but indescribable. They freely set relationship of geometry and space between abstract and concrete, and then aroused audience's infinite imagination. Just similar attitudes to Hadid's architecture drawing collections, every professional agreed that Liebskind's geometric attractions were just a game of mathematics and paintings. However, Liebskind finally proved his extraordinary design talent with facts. People started to accept another different interpretation of modern architecture when the Jewish Memorial Museum was finally built and immediately became new landmark of Berlin. In today's world, there is another world leading master, a contemporary designer of Spain, Calatrava, who inherited genius of Leonardo Da Vinci and Gaudi and obtained a new understanding of *abstraction & geometry* from nature, and created a series of artistic structures between sculpture and architecture. Calatrava was an outstanding engineer architect, artist, inventor, and scientist with his doctorate in architecture and engineering. He followed example of Leonardo Da Vinci and was curious to nature. He was influenced by Gaudi, an architecture predecessor, so deep that he was interested in human skeleton structure and its operation principle. His engineering training enabled him to build his own understanding patterns of abstract thinking and geometric dimension through logical system, and thus to establish his own unique architectural creation system. All of his architecture and bridge works became local attractions and landmarks.

Although Chinese reform and opening-up provided opportunities to comprehensively contact with their concepts and works of every famous architect above and national huge construction projects provided us stages for Chinese architects to endlessly show their originality, Chinese contemporary architects are still exploring. What do Chinese architects still lack? We surely expect better owners, more reasonable building systems, mature design education systems and so on. However, what we need most is to calm down and to think carefully and systematically about our own design methods and creative concept in order to basically build a real understanding of *abstraction & geometry* thinking pattern. And during process of establishing the thinking pattern, we elaborately examine our national cultural background, national symbols and traditional design wisdom. Meanwhile, we, like many senior masters, love nature, observe nature, and regard human design heritage as our own soul treasures. For this purpose, we will expect better future.

4. Formation of the Finnish School of Architecture

Eliel Saarinen was the inaugurator of the Finnish school of architecture and the founder of the Nordic school of design. Although he spent most of his later life in the United States, as one of the top American modern architecture masters and one of the most important founders of American modern design-and-art education, his design practice in Finland in the first half life still had irreplaceable influence. His son, Eero Saarinen, was one of the most talented architects in the 20th century. However, he was regarded as an extraneous member of the Finnish school of architec-ure because his design career was mostly in the United tates of America. Eliel Saarinen was the flag-bearer of he Finnish national romantic architecture movement in he late of the 19th and the early 20th centuries. Although inland is geographically far from the core of European ulture and art at that time such as France, Belgium, the Jnited Kingdom, Austria and so on, lots of Finnish ar-hitects and artists, represented by Eliel Saarinen, still imely introduced the most fashionable Arts and Crafts Aovement and Art Nouveau Movement from Europe into inland. Combined with their own national culture tradi-ions, they created a large amount of architectural classics vith Finnish national romantic style. The classic works f Eliel Saarinen and other Finnish architects, such as Ielsinki Central Railway Station, Finnish National Mu-eum, and a large number of administration buildings nd apartments, were all masterpieces in the architecture istory, even if we measure them with today's vision and tandards. The buildings carefully inherited the Europe-n classical architecture traditions, and also integrated nique cultural elements of Finland and northern Europe. Based on the consistent *abstraction & geometry* thinking attern, Eliel Saarinen integrated the European classical olumn styles and decoration theme from Nordic natural lements, and created a series of peaceful, warm, ecolog-cal, and healthy buildings with people- oriented and de-igned spaces with livability and usability, which initiated he Nordic school of humanistic functionalism.

Alvar Aalto, hailed by a famous architectural historian, Gideon, as a grandmaster of modern architecture and nodern design in one of the five masters of modern ar-hitecture, realized Finnish School of architecture well nown throughout the world and began to widely impact global architects. Compared with other four modern archi-ecture founders, Wright, Gropius, Ludwig Mies Van der Rohe and Le Corbusier, Aalto was comparatively young nd had relatively less creative works. However, Aalto was he top master of integrated design. Meanwhile, he was lso an architecture ideologist with profound and lasting nfluence, paying more attention to society, human beings, nvironmental ecology and sustainable architecture de-elopment. Therefore, the Finnish school of architec-ure had different thinking characteristics from other arly modernism architecture and also maintained a trong influence during the whole 20th century with its riginal ideas and a consistent attention to ecological nvironment. Aalto was praised as a "genius" by Wright ecause Aalto was able to integrate the classical column tyles of ancient Greece and ancient Rome with pioneer deas of his contemporary colleagues, and then began to reate his own Finnish-style design language through mazing creative approaches. These design languag-s embraced mathematical grammar and vocabulary of Euclidean geometry and Nordic geometry in ways of comprehensive collection and selective absorption. Then they, based on local building materials in Finland as their core, constructed an abstract space system serv-ing basic functions but surpassing basic functional re-quirements. Aalto's architecture was diverse but never boring. Every time staying in his buildings, users had their admiration from deep hearts. Aalto's architectural talent is also manifested in the pioneering research of building prototypes. He also had pioneering contribu-tions in the research and design of sanatoriums, hospi-tals, libraries, theatres, universities, churches, houses and architectural types. In addition, Aalto is also one of the most important pioneering masters in the fields of modern furniture and industrial design, who has revo-lutionary influences on the creative concept of modern design.

It is well known that geniuses are congenital and cannot be copied or are difficult to be replicated. Modern archi-tectural geniuses, such as Wright, Corbusier and Aalto, started a new design style, but it is difficult to have such a generation of masters again. Hence, if great architectural schools hope to continuously create excellent architects, they must seriously consider the issues of architectural re-search and design education. In this respect, the Finnish school of architecture was fortunate to own not only Aalto, but also Blomstedt.

The present author systematically introduced two archi-tectural tutors, Aalto and Blomstedt, in an earlier paper *Two Poles of Finnish Architecture,* who had the greatest influences on Finnish architecture in the 20th century. Aal-to's influence was mainly reflected in setting an example with his gifted achievements and establishing cultural self-confidence. However, Aalto spent only a short time in teaching. He relied on his works, but did not form phil-osophical concept system. Nevertheless, Blomstedt was contrary to Aalto in many aspects and complemented him. Although he often carried out design practices by follow-ing the traditions of Finnish architecture professors, he spent most of his time in systematic teaching and research on architectural scales and harmonious dimensions. Blom-stedt was born in a learned family and was well aware of the significance of architectural education to Finnish sus-tainable architecture development. In the meantime, he devoted his whole life to the research theme about advo-cating what kind of architectural teaching.

Juhani Pallasmaa, a Finnish contemporary architect and a world-renowned architectural educator and critic, told the author that there were two gold mines, Aalto and Blomstedt, that seemed never to be excavated among the Finnish modern architectural treasures. Blomstedt was endowed with extraordinary talent, such as profound thought and excellent insight, who had a long-term effect on future generations. As he says: "You must understand the oldest if you want to achieve the most modern achievement." When the author wrote relevant articles in the early years, he had visited the archive in Finnish Architecture Museum several times to pry into mysteries of Blomstedt as a gold mine. And he found that the depth and breadth of Blomstedt's basic research in his life-time work can, to some extent, be compared with Aalto's design achievements. Le Corbusier and Blomstedt were the two most successful design tutors in architectural modular study of the 20th century. The former did research together with a large number of successful design practices, who influenced the global architecture. The latter started from discussing music and created a new style of Finnish architecture which became the driving force to help the Finnish school of architecture to still exist. In the archive of Finnish Architecture Museum, the author was lucky to be able to witness Blomstedt's original drawings of proportion, scale and harmonious relations in his modular system research. Blomstedt drew dozens of color analysis charts based on architectural scale and human proportion system of ancient Egypt. He carefully sorted out the column style research manuscripts of the masters from Alberti to Palladio during the Italian renaissance. He analyzed and drew the music analysis diagrams of famous musicians in the European Baroque period. Since then, the author began to really understand the strength and warmth of Finnish modern architecture. The author also deeply realized the great power of in-depth research while recalling Leonardo Da Vinci's notes and manuscripts, Craig's drawing research manuscripts and Picasso's manuscripts of human proportion research in the relevant European exhibitions during the past few years. The author, once again, recalled the remarks of his supervisor, Professor Guo Husheng, about academic research: "no breadth then no depth".

Blomstedt's modular research was extensive and deep with profound effect, and his design practices were exquisite and meticulous, which were regarded as excellent examples. Thus, architectural education, based on this kind of research and practice, can entirely reflect its effectiveness, strength and temperature. In other words, the influence of Aalto was inspirational while Blomstedt's was systematic in Finnish architecture. The former was optional and difficult to inherit while the latter had strong sense of inheritance through systematic teaching and research. Consequently, it can be found that there were few Finnish architects who can inherit Aalto's inspiration, but most of them were nurtured and trained by Blomstedt's teaching system. Blomstedt's architectural education powerfully established the mainstream of functionalism and humanism in the Finnish school of architecture. And the diversified ecology of Finnish architecture was still found befor and after his passing, which also, from another perspec tive, showed the beneficial influence of his teaching.

In the golden age of Finnish architecture with two mas ters, Aalto and Blomstedt, a group of Finnish architect with great creative potentials had emerged alread Among them, the most prominent architects were Piet la and Aarno Ruusuvuori. Their common features wer equal emphases on teaching, research and design practice Pietila had more or less signs of Aalto's genius and inspi ration, while Ruusuvuori, an assistant of Blomstedt for long time, firmly inherited the humanistic functionalism theme of Finnish architecture. Pietila, unlike Aalto, wa fond of writing and research, and especially liked creat ing architectural masterpieces with different styles base on totally different cultural inspirations. However, Ruu suvuori, to a great extent, followed Blomstedt's researc and teaching methods, but in design practice he devel oped a completely different concrete building style fron his tutor. His architecture was rigorous but powerfu and it was simultaneously filled with emotion and vitalit which showed thoroughly the power of Finnish modula research. Actually, Ruusuvuori's masterpieces of concret building in the 1960s not only started a new era of Finn ish architecture, but also had a great influence on a larg number of foreign architects, including Tadao Ando, i which the power of *abstraction & geometry* thinking pat tern was its fundamental reason of broad influence.

After Pietila and Ruusuvuori, the Finnish school of archi tecture began to show a trend of diversification, amon which there were the Finnish local style represented b Juha Leiviska who introduced design concept with musi modulus, the Finnish high-tech style by Helin as its repre sentative who echoed international high-tech school, th Finnish personalized style represented by Pekka Salmin en who applied material research to guide design creativi ty, the style of Finnish steel structure represented by Es Piironen, and the Finnish scholar design style directed b Juhani Pallasmaa as its academic research representa tive, etc. Nevertheless, all styles, to a great extent, firml supported the creative thinking concept of *abstraction & geometry*. Actually, ultimate insistence on *abstraction & geometry* was regarded as the foundation of most Finnis architects, such as Heikkinen-Komonen Architects, on famous example.

5. Heikkinen-Komonen Architects: From *Abstraction & Geometry* t Architecture Masterpieces

Mikko Heikkinen and Markku Komonen were both bor in the 1940s and graduated from Helsinki University o Technology (now Aalto University) in the mid-1970s. The grew up at the age of the last stage when Finnish architec ure was dominated by Aalto. The development of Euro-ean modern architecture was initiated by Germany and rance, which was famous for its international style full f coldness and coolness until Aalto took over modernism nd established a humanistic style of modern architecture hrough the concept of regionalism and people–orienta-ion. Thus, Finland was promoted to the leading position f modern architecture movement. However, along with huge number of housing issues caused by economic evelopment and the accelerating demand of high-tech n construction, personalized factors expressed by Aalto vere questioned in Finland. Blomstedt and Ruusuvuori, n questions and disputes, began to set up an architectural oncept of new rationalism entirely based on *abstraction & geometry* thinking pattern. Simultaneously, Pietila tarted from Aalto's concept of regionalism and estab-shed a design concept of construction bionomics based n his rich theory and practice. When Aalto became world amous, his dominant position in his motherland was chal-enged in several ways, which was exactly the important eason why Finnish architecture continuously kept on im-rovement and maintained a high level in the world. That s to say, a vibrant culture would not be influenced by only ne thought for a long time and a culture would soon de-ay and die if there were no contradictions and challenges.

Vhen Heikkinen and Komonen began to work, the overall rchitecture styles of Finland were in a period of diversi-cation, which were called by domestic and foreign critics s regionalism, humanism, romanticism, romantic mod-rnism, new rationalism, ecologism and so on. Heikkinen nd Komonen initiated their own design concept in this ackground and eventually turned to humanistic func-ionalism.

Ieikkinen and Komonen were famous for the Finnish Sci-nce Center, which won the National Architectural Com-etition in 1986 and officially opened to public in 1989. Ieikkinen and Komonen became famous overnight and imultaneously set up a rigorous spirit of science explo-ation and a working attitude of refined details during he project design process. This project confronted great hallenges at that time, not only because it contained ew ideas, but also its architects, at all stages of design, vere required to repeatedly study mathematics, optics, coustics, electricity, biology, chemistry, astronomy, geog-aphy, medicine and natural sciences. Hence, they firmly elieved that architecture was a holistic design consid-ring the contradiction and coordination of nature in a nified way and architects needed to search regular and alanced forces from natural chaos and refine them, just imilar to science research. Heikkinen appreciated greatly motto of Paul Valerie, a French philosopher, "there are wo things, disorder and order, that never stop interfering vith the world." Disorder and order in human being's heart exist everywhere in nature. There are conflict and harmony between premonition and randomness. As far as architecture design is concerned, this attitude from edg-es of science research contains actually infinite aesthetic potentials. Disorder needs abstraction and examination, while order depends on geometric model to interpret, thus a design system is established. Nature itself is a balanced and beautiful holistic entirety. Nature, within the limited thinking range of human beings, is fantastic and harmo-nious, like a flash of lightning and a strip of rainbow after rain and sunrise or sunset of everyday which would pro-vide irresistible influences on being's aesthetic thinking. Heikkinen and Komonen started from their first well-known work, Finnish Science Center, and immediately fell into a deep fascination with *abstraction & geometry* thinking pattern and naturally continued to their subse-quent projects.

To Heikkinen and Komonen, scientific thinking based on *abstraction & geometry* would not be boring, but can be unpredictable and full of interesting. Therefore, their ar-chitecture works always had both seriousness and sense of humor, changing between temperance and casualness. Full understanding of *abstraction & geometry* at any level can actually produce endless changeable patterns in archi-tectural space creation. The functional sequence of Finn-ish Science Center was composed of different elements in solid geometry, in which different nodes of various mate-rials were introduced to connect structures where cubes and spheres in pure geometry took on different functional spaces in various variants as an interpretation of scientific spatial morphology.

Heikkinen and Komonen had a belief that architecture was not pursuit of style, but about art of poetry. They firmly believed that the persistence of *abstraction and ge-ometry* thinking is a kind of poetry and the main theme of architecture design after experiencing their school days and later design practices, which ran through every ar-chitecture work of Heikkinen-Komonen Architects. The intentional construction of geometric volume in their architectures was often refined to extreme, the same as paintings and installation art of Kandinsky, Malevich and Moholy Nagy, highlighting poetic presentation. The cube combination of Rovaniemi Airport Terminal Building in Finland intended to express eternal atmosphere of a cold climate by metal boxes bearing strong geometric charac-teristics, and also to exhibit the industrial aesthetics of minimalism. The Kuopio Emergency College of Finland reflected the concept of training people to calmly face fire and danger in building overall layout. Among cold geomet-ric form combination, the long and narrow classrooms compare sharply with crescent-shaped dormitories. Heik-kinen and Komonen perfectly solved all functional prob-lems in the European Film School in Denmark through the combination of pure geometric bodies. Two mature Finnish architects, in the project of Finnish Embassy in Washington, the United States of America, interpreted poetic quality with geometric forms by the most fashionable metals and glasses. Cubic solemnity of its facade and enthusiasm of interior metal staircase seemed to be a conservative and silent Finnish who politely concealed his inside passions. At the opening ceremony of this embassy in 1994, Ieoh Ming Pei himself specially went to express his congratulations. Rigorous and specific geometries and coordination with different site environments constituted the important cornerstone of Heikkinen-Komonen's works, which were like huge concrete semi-circular arc in Helsinki Cultural Center containing warmly a library's cultural activity center. Among their completed projects in recent years, such as Center for Systems Biology in Dresden of Germany, Schönbühl Park of Lutzern, Switzerland, Savonlinna Library of Finland, etc, Heikkinen and Komonen integrated more high-tech achievements into their designs and actively participated in research and development of new materials and building accessories. This kind of research and development was very often reflected in their frequent and precise model makings and each of their design projects was a kind of scientific discussion to themselves. Heikkinen and Komonen repeatedly applied various combinations of straight lines and arcs, cubes and spheres accompanying with various types of metal screens, different kinds of glasses and other industrial products as an abstract game of formalism on the surface of their designed buildings. But in essence, what they always pursue was to achieve a kind of poetry in the exploration of pure form ideality, and to complete the humanistic functional design concept full of idealism in their design works.

The development of western architecture, from ancient Egypt, ancient West Asia to ancient Greece and ancient Rome, went through the Middle? Ages to the Renaissance, then experienced Baroque, Rococo, Neoclassicism, Arts and Crafts Movement, Art Nouveau Movement, and New Decoration Movement, and finally came into modernism. Its unchanged theme was to pursue the larger and more humanized spaces and its eternal ideality was to show a poetic expression with abstract thinking and geometric dimension. These laureate poets of modernist architecture, from Wright and Eliel Saarinen to Le Corbusier, to Walte Gropius, Ludwig Mies Van der Rohe, Aalto and Marc Breuer, to Louis Isadore Kahn, Ieoh Ming Pei, Blomsted Pietila and Ruusuvuori, composed some beautiful chap ters of modern architecture with their extraordinary ta ents one by one. *Abstraction & geometry*, in colorful chap ters, was still the eternal theme of their creative desigr How to stimulate the creative poetry in the new era? Hov to maintain and promote the core of humanistic function alism in the modern architectural movement at the globa information age? Being as both architects and professor of architecture colleges, Heikkinen and Komonen were al ways consistent in thinking about these above problem and attempting to give some meaningful answers throug their persistent design practice. They, based on the cre ative pattern of *abstraction & geometry*, will forever ex plore the potential of modern architecture with a very per sistent spirit, that is, to what extent human architectur can explain the relationship between humans and nature On the one hand, their works were in almost enthusiasti pursuit of pure geometry, and on the other hand, they ful ly accepted all new materials and technologies. The com bination of both aspects naturally formed an optimisti attitude towards human beings, environment and future

The human society, in the past fifty years, made grea changes, and the development of science and technolog in our era was astonishing from macro to micro levels But the so-called high-tech in architecture still remaine at the level of steam engine era. Architects should not b easily satisfied but should pursue and perfect the creativ pattern of *abstraction & geometry* with a more persisten attitude and pay close attention to the progress of scienc and technology and the development of ecology. Only i this way can they achieve the poetic creation and aestheti sublimation. Heikkinen-Komonen Architects' works ha shown a kind of solid sense of beauty, which came from ab stract simplicity and precise focus on geometric form lan guage. Their works were full of creative thinking, whic included conscious and unconscious absorption of variou cultural trend, and a wise appreciation and experienc of natural environment, scientific theory, architectura structure and material details.

附录 2 | Appendix 2 “模型草案”英原文

SKETCHING WITH MODELS

Mikko Heikkinen

In the old days, your career in an architect's office in Finland used to start as a junior assistant building architectural scale models. At my very first job I soon found myself cutting cardboard with a Stanley knife. Of course, I had previously built airplane models as well as landscaping made of gypsum for my own model railways, but architectural models were terra incognita to me. The resistance of matter was overwhelming, it was hard to make straight cuts through balsa sticks and glue spread out from the joints. The facade of the model I was working with was supposed to be a kind of modular scaffolding made of thin metal rods, which was fashionable that time. I ordered the rods from a small workshop located in the yard close to our office. Those rods were probably the only precise elements in my structure. I then had the honour of taking my “oeuvre” to the meeting where the decisions about the project were to be made. The place was no less than the Restaurant Savoy in Helsinki designed by Alvar Aalto. Naturally, I had heard about this masterpiece by the legendary architect, and I eagerly waited to be finally able to see it in real life. At the door, an unfriendly porter scanned my tennis shoes and worn U.S. Army jacket, thanked me, took the model from my hands and shut the door.

Since that time, the everyday life at architects' studios has completely changed. Work takes place digitally on the computer screen and ever more realistic 3D-modelling software challenges even photographic reality. However, the evolution of both hardware and software does not seem to have superseded traditional working with architectural models. New materials and machinery, such as styrofoam, laser cutting and 3D printing, have created new possibilities.

Many well-known architects have admitted that models have been essential tools in their design process even already in the earliest phase of projects. Architect Peter Zumthor demonstrates his thoughts with images made on top of photos of models. Large architectural models dominate his studio and discussions with his assistants, and presentations to his clients take place around them. These objects are characterized by an inventive and often surprising palette of materials, which make them look like artworks in themselves.

Japanese architect Sou Fujimoto and his design team communicate with each other through models. And models are made in abundance: for the Serpentine Gallery Pavilion in London in 2014 there was almost 70 of them. Sometimes, after one hundred versions, they might start from scratch. He has stated, for instance, that “*for the very early phases, conceptual sketch models visualize what we are thinking, or what could be the topic, or simply what the possible volume is...*”

For the Portuguese architect duo Aires Mateus: “*The interest in models is a phenomenological obsession.*” At their firm, they strive to build large-scale models already in the beginning of a project. They see that the danger with miniature-size models is that even an ugly solution may look nice: “*We like to design our architecture as if we were looking at it from within, which is why it's quite normal for our models to be on a 1:20 scale.*”

At the SANAA studio in Tokyo work on a project in the beginning is *model-centric.* Everybody in the design team works with models, often without any drawings, and discussions take place around models. Ryue Nishizawa thinks that “*looking at the model, you may think this idea is no good, or this one is better than I thought. It opens the way to the next step.”*

The conceptual models, or rather the material studies, presented in the catalogue for the exhibition *Natural History* on the works by architects Jacques Herzog and Pierre de Meuron demonstrate how a change in matter can completely change the character of the form. The transparency, manufacture, precision of form, and texture of the model may all direct the design process.

At sometime at the end of the 1990s, I visited Steven Holl's office on Hudson Street in New York. The wall of the reception space was covered entirely with palm-sized model studies for the Helsinki Museum of Contemporary Art competition from 1992. At first glance, they did not seem to have anything in common, but after a second look it was possible to trace a line of development towards what would become the winning entry in the competition.

Many of the models face a sudden and deserved death following a “review and reject” process and end up in the trash bin under the worktable. Some survive as relics in the evolution of the design process. A model might contain a new solution or a new approach for an unknown future project, and could be reused in a completely different context. In Bjarke Ingels' office BIG all old models are stored as an ideas bank for possible new use.

In both a storage and exhibition vitrine, models demonstrate the persistent efforts by their creators to find a better solution. In the catalogue for their exhibition *Natural History* Jacques Herzog and Pierre de Meuron raise the question of whether a model is architecture in itself or just waste left from the design process, and give the response: *“These archived objects are therefore nothing but waste products, since the immaterial, mental processes of understanding, learning, and developing always have priority.”*

For a long time models in our office were mostly only built for the final presentation of a completed design, either in an architectural competition or for a client. Sketch designing took place mainly with pen and paper. After the planning process shifted from open worktables to personal computer screens, the projects' material presence in the office disappeared. By using rendering programs, it is possible to create 3D models and to rotate them on the computer screen, but the image is nevertheless still two-dimensional, and looks the same to everyone looking at it. Starting to sketch with models seemed to compensate for the missing link, that is, the social focus within the team. Through a real scale model, the progress of a project takes on a physical, concrete form, and the whole design team is able to refer to it, either for inspiration or as a challenge to excel. According to Ryue Nishizawa of SANAA: *“If you have an idea, make a model. It turns the idea into something in the outside world, something you can see.”*

For me personally, sketch designing with a model, an thinking with the help of a model, have often replace sketching by drawing. At the early stages of a project, th use of a pencil may have been reduced simply to preparin the designs of models. Building a model always already requires some kind of vision, that is, a concept for solvin the design task. On the other hand, a sketch model force you to transform the idea into a viable and demonstrabl form. It is certainly difficult to concretize unclear thinkin with a model. Sometimes the model reveals better tha words what I am thinking or what I didn't know I intended.

Our proposal in the architectural competition for th Finnish Pavilion for Expo 2010 in Shanghai, China, ha its origins in the stump of a felled tree situated in a par near our office. I then searched the forest for a felled birc tree of a suitable size, from which I sawed a thin disk. B carving it into the desired shape and covering the cut surfaces with mirrored film, the piece of wood showed at single glance the concept of a forest-like exterior, inside which is revealed a modern exhibition pavilion.

In the architectural competition for the Finnish Embass in Tokyo, Japan, we wanted to “fuse” into a single form the embassy building and a residential building that wa to be built on the same plot. By folding thin tracing paper, it was easy to create sculptural and almost seamles studies of the massing, the overall form of which could b emphasized by lighting them from below. The object-lik quality often brought out by the small size of the sketc model and the bird's eye perspective is corrected whe photographed. A two-dimensional photo of a 3D mode generates an illusion of the building in full scale.

The contents on the shelves in a craft supplies store already in themselves inspire me: “I wonder what could b done with that one there?” The most interesting materials can be found anywhere, simply by accident. One o our model builders says he gets inspired in junkyards, fro where he has acquired supplies for possible future use. Th same form can take on a completely different character wit a different choice of materials. A 3D printer is an excellen tool in producing models that show the massing and complex components, but it gives an unresponsive impressio of the material and its properties, for instance differen degrees of transparency and reflectivity.

Using collage techniques, it is easy in a model to conve white cardboard into, say, ceramic-patterned concrete. I the Flooranaukio housing project in Helsinki the idea o using waste porcelain from the adjacent Arabia ceramic factory in the facades of the building led to experiment ions in which ceramic textures scanned from the internet vere glued onto the surface of a 1:200 scale model cut in ccordance with the decorative pattern on a vase produced y the factory in the 1930s.

model can sometimes prove to be the quickest way of emonstrating the basic idea in a proposal. During the deign process for the Alma Media headquarters in Helsinki we ad just one night to come up with a solution for a particular esign problem. The plastic massing of the building, which ollows the edges of the plot as well as the orientations of the urrounding buildings, was depicted using acrylic boards. he latter represented the different floor levels, which were unched through with wooden dowels representing stairs. he form of the building – the strip-like relationship of the olid surfaces and openings of the facade – was realised hrough overhead projector transparencies, onto which he various brand names of the media building were coped. Here, too, an illusion of the real character of the building was created through photos of the scale model.

or the competition proposal for the Perhelä housing block Järvenpää we studied, with the help of a scale model, how he impression of a building consisting of vertical compoents changes, depending on the viewing angle and lighting. series of photos taken of the scale model gave the impression of a real building to an object that resembled miniature sculpture, and these were presented in the competition display panels as a 360-degree view around the block.

Sometimes, however, the model works best at its own scale – turning it around in the palm of your hand, like an object akin to an amulet which is meant to be touched. The conceptual model for the Dresden Centre for Systems Biology was made in solid mahogany, representing the concrete labyrinthine core at the heart of the building. The model resembles a Japanese Kumiki puzzle. Sitting in the palm of the hand, it gave not only a visual sensation but also a haptic one, due to the indentations in the fist-sized object.

This book presents a collection of models made at our office during the early phases of architectural competitions or commissions. Most of them were built beside the computer screen or drawing board with the basic tools and materials at hand. Models for the final presentations, however, have been ordered from external professional workshops. Model building has supported the work of the rest of the design team and vice versa, with the design process providing feedback for the development of the models. Most conceptual models have, just like any project's design sketches, disappeared as paper waste, yet photos of them have remained as memory traces of the different stages of the design process.

附录 3 | Appendix 3 相关研究成果与学术活动

主要作品

系统生物学中心，德国，德雷斯顿，2017 年；
坎加沙拉艺术中心，芬兰，坎加沙拉，2015 年；
斯托布尔公园，瑞士，卢塞恩，2014 年；
桑塔哈米纳住宅区，芬兰，赫尔辛基，2015 年；
萨翁林纳主图书馆，芬兰，2013 年；
弗洛拉诺基奥住宅区，芬兰，赫尔辛基，2012 年；
海门林纳省档案馆，芬兰，海门林纳，2009 年；
应急服务学院，芬兰，库奥皮奥，1992 年、1994 年、2005 年；
拉彭兰塔技术大学第七期工程，芬兰，拉彭兰塔，2004 年；
参议院办公大楼与国家福利与健康研究中心，芬兰，赫尔辛基，2002 年；
马克斯普朗克分子生物学和遗传学研究院，德国，德雷斯顿，2001 年；
沃塔洛文化中心，芬兰，赫尔辛基，2001 年；
科斯克图斯预制房屋，芬兰，图苏拉，2000 年；
卢米媒体中心，芬兰，赫尔辛基，1999 年；
西非项目，几内亚，1994—1999 年；
卡勒瓦拉和卡雷利安文化信息中心（尤米科考），芬兰，库赫莫，1999 年；
麦当劳总部大楼，芬兰，赫尔辛基，1997 年；
老年住房和福利中心，芬兰，万塔，1994 年；
芬兰驻美大使馆，美国，华盛顿特区，1994 年；
欧洲电影学院，丹麦，埃贝尔托夫特，1993 年；
罗瓦涅米机场航站楼，芬兰，罗瓦涅米，1992 年、2001 年；
赫尤里卡科学中心，芬兰，万塔，1989 年、2017 年；
洛马尼梅尔山度假中心，芬兰，艾内科斯基，1987 年。

Major works include

Center for Systems Biology, Dresden, Germany, 2017
Kangasala Arts Centre, Kangasala, Finland, 2015
Schönbühl Park, Lucerne, Switzerland, 2014
Santahamina House, Helsinki, Finland, 2015
Savonlinna Main Library, Savonlinna, Finland, 2013
Flooranaukio Housing Block, Helsinki 2012
Hämeenlinna Provincial Archive, Hämeenlinna, Finland, 2009
Emergency Services College, Kuopio, Finland, 1992, 1994, 2005
Lappeenranta University of Technology, phase 7, Lappeenranta, Finland, 2004
Senate Properties and STAKES Office Buildings, Helsinki, Finland, 200
Max Planck Institute for Molecular Biology and Genetics, Dresden, Germany, 2001
Cultural Centre *Vuotalo*, Helsinki, Finland, 2001
Prefabricated House *Kosketus*, Tuusula etc, Finland, 2000
Media Centre *Lume*, Helsinki, Finland, 1999
Projects in West Africa, Guinea, 1994-1999
Information Centre *Juminkeko* for the *Kalevala* and Karelian Cultur Kuhmo, 1999
McDonald's Headquarters, Helsinki, Finland, 1997
Senior Citizen Housing and Amenity Centre, Vantaa, Finland, 1994
Finnish Embassy, Washington D.C., U.S.A., 1994
European Film College, Ebeltoft, Denmark, 1993
Rovaniemi Airport Terminal, Rovaniemi, Finland, 1992, 2001
Finnish Science Centre *Heureka*, Vantaa, Finland, 1989, 2017
Vacation Hotel *Lomaniemelä*, Äänekoski, Finland, 1987

专著

《芬兰驻华盛顿大使馆》，尤卡 · 沃特撒里作序，克里斯蒂纳 · 海伦纽斯、塞韦里 · 布隆斯泰特、简 · C · 洛菲勒和图拉 · 伊恩 - 考斯科那撰写，芬兰，霍洛拉，2005 年；
《伟大建筑师的优秀作品集：海基宁—科莫宁》，中国，中国建筑工业出版社，2003 年；
《海基宁—科莫宁》，方海编，尤哈尼 · 帕拉斯玛作序，中国建筑工业出版社，中国，北京，2002 年；
《温柔的桥梁：建筑、艺术和科学》，R · 安东尼 · 海曼、格哈德 · 马克、尤哈尼 · 帕拉斯玛和马里诺 · 泽瑞尔撰写，博克豪斯出版社，瑞士，巴塞尔，2002 年；
《海基宁—科莫宁》，威廉 · 摩根编，尤哈尼 · 帕拉斯玛作序，莫纳塞利出版社，美国，纽约，2000 年；
《韩国建筑师》，彼得 · 马肯斯作序，1995 年 6 月；
《海基宁—科莫宁》，彼得 · 戴维作序，古斯塔沃 · 吉利出版社，西班牙，巴塞罗那，1994 年。

Monographs

The Embassy of Finland in Washington D.C., Hollola 2005, foreword b Jukka Valtasaari, articles by Kristiina Helenius, Severi Blomstedt, Jan C. Loeffler and Tuula Yrjö-Koskinen
The Excellent Works of the Great Architects: Heikkinen - Komonen China 2003
Heikkinen - Komonen, China Architecture & Building Press, Chin 2002, edited by Fang Hai, introduction by Juhani Pallasmaa
Gentle Bridges: Architecture, Art and Science, articles by R. Anthon Hyman, Gerhard Mack, Juhani Pallasmaa and Marino Zerial, Birkhäuser, Basel 2002
Heikkinen + Komonen, The Monacelli Press, New York 2000, edited b William Morgan, introduction by Juhani Pallasmaa
Korean Architects, 1995/6, introduction by Peter MacKeith
Heikkinen & Komonen, Gustavo Gili, Barcelona 1994, introduction b Peter Davey

建筑与设计类书籍的报道

2014 年

《设计材料》，维多利亚·巴拉德·贝尔和帕特里夏·兰德著，普林斯顿建筑出版社，美国，纽约，第 56 ~ 57 页，海门林纳省档案馆。

2013 年

《致电并回复乔治·施泰曼因》，图恩艺术博物馆，谢德格和施皮斯，第 57 ~ 72 页，科米。增多的雕塑（1997—2007 年），元数据逻辑（1999—2000 年）。

2012 年

《邂逅 2，尤哈尼·帕拉斯马建筑散文》，载于《禅与建筑艺术——海基宁—科莫宁建筑事务所作品集》，第 152 ~ 162 页，2012 年；
《芬兰杰出建筑》，卡里帕西拉·贾塔尔·亚努米，芬兰，马亨基，第 40 ~ 50 页，弗洛拉诺基奥住宅区设计，2013 年；
《现代斯堪的纳维亚建筑——北欧之光》，亨利·普卢默著，泰晤士和哈德生出版社，英国，伦敦，2012 年；罗瓦涅米机场，第 81 页，麦当劳总部，第 124 页；
《桑科鲁瓦》，200 年前的城市建筑监理，玛亚—赫肯拉·考皮宁，赫尔辛基建筑控制系，哈沃特—阿瑞纳出版社，第 156 页，2012 年；
《2011 年竞赛年册》，竞赛项目公司出版，第 30 ~ 33 页，瑟拉格斯博物馆，2012 年；
《建筑立面材料语言像素墙》，凤凰出版传媒公司出版，中国，江苏，2012 年，海门林纳省档案馆；
《鲁克塔米森·鲁祖 2007—2011 年》，赫尔辛基市建筑控制委员会，第 21 页，弗洛拉诺基奥住宅区设计，2012 年。

2011 年

《100×N 建筑形态与外立面》，香港建筑出版社，中国，香港，2011 年，海门林纳省档案馆；
《美—实用—耐久性：两个世纪大厦的状态》，埃迪塔著，芬兰，波尔沃，2011 年，应急服务学院，第四期，第 76 ~ 79 页，芬兰大使馆，第 91 ~ 93 页，参议院办公大楼与国家福利与健康研究中心，第 106 ~ 109 页，海门林纳省档案馆，第 142 ~ 145 页；
《工业场地再开发设计：建筑师、规划师和开发者指南》，卡罗尔·贝伦斯著，约翰威利出版公司，美国，新泽西州，霍博肯市，2011 年。

2010 年

《建筑亮点（第三卷）》，上林有限公司，中国，香港，2010 年；应急服务学院，第四期；拉彭兰塔理工大学，第七期，第 218 ~ 223 页；
《芬兰当代建筑》，安东内洛·阿里西著，24 ORE 文化公司出版，意大利，米兰，2010 年；罗瓦涅米机场，第 28 ~ 31 页，拉彭兰塔理工大学，第七期，第 114 ~ 117 页；
《紧急和临时建筑手册》，M·科拉多编，西斯塔米出版，2010 年。

Architecture and Design Books

2014

Materials for Design, Victoria Ballard Bell and Patricia Rand, Princeton Architectural Press, New York, p56-57, Hämeenlinna Provincial Archive

2013

Call and Response George Steinmann im Dialog, Kunstmuseum Thun Scheidegger & Spiess, p57-72, Komi. A Growing Sculpture (1997-2007), Metalog (1999/2000)

2012

Encounters 2, Juhani Pallasmaa-Architectural Essays, p152-162 "Zen and the Art of Making Architecture-The Work of Heikkinen-Komonen"
Finnish Architecture with Edge, Kari Palsila ja Tarja Nurmi, Maahenki 2013, p40-50 Flooranaukio Housing
Modern Scandinavian Architecture *Nordic Light*, Henry Plummer, Thames & Hudson 2012, p81 Rovaniemi Airport, p124 Mc Donald's Hq
Saanko luvan, 200 vuotta pääkaupungin rakennusvalvontaa, Marja-Heikkilä Kauppinen, Helsinki Building Control Department 2012, p156 Hartwall-areena
2011 Competitions Annual, the Competitions Project Inc 2012, p30-33 Serlachius-museum
Wall Elements, Pixel Wall, Phoenix Publishing & Media Inc 2012, China, Hämeenlinna Provincial Archive
Rakentamisen ruusu 2007-2011, Helsinki City Building Control Comission 2012, p21 Flooranaukio Housing

2011

100×N Architectural Shape and Skin, Hong Kong Architecture Press, Hong Kong 2011, Hämeenlinna Provincial Archive
Kauneus-käytännöllisyys-kestävyys, Valtion rakentamisen kaksi vuosisataa, Edita, Porvoo 2011, p76-79 Emergency Services College phase IV, p91-93 Embassy of Finland, p106-109 Stakes and Senate properties, p142-145 Hämeenlinna Provincial Archive
Redeveloping Industrial Sites, A Guide for Architects, Planners and Developers, Carol Berens, John Wiley & Sons, Inc. Hoboken, New Jersey 2011

2010

Architecture Highlights vol. 3, Shanglin A&C Limited, Hong Kong 2010, p218-223 Emergency Services College phase IV and Lappeenranta University of Technology phase VII
Contemporary Architecture Finland, Antonello Alici, 24 ORE Cultura srl, Milan 2010, p28-31 Rovaniemi Airport, p114-117 Lappeenranta University of Technology phase VII
Manuale di Architettura di emergenza e temporanea, edited by M. Corrado, sistemi editoriali 2010

2009 年

《文化中心：建筑 1990—2011》，塞西莉亚·比奥内、莫塔著，意大利，米兰，2009 年；卡雷利安文化信息中心（尤米科考），芬兰库赫莫，第 104 ～ 109 页；
《4×4 建筑十字路口》，马尔科·D·安纳蒂斯、萨拉·西波列蒂和考得利贝特工作室，阿斯科利皮切诺建筑学院，2009 年；
《芬兰的混凝土建筑》，SBK—撒阿奥，约瓦斯卡拉，2009 年，芬兰科学中心，第 182 ～ 183 页。

2008 年

《北欧建筑师写作》，迈克尔·亚斯格特·安徒生编，劳特利奇出版社，英国，阿宾顿，2008 年，《马尔库·科莫宁：建筑、技术和艺术》，第 183 ～ 188 页；
《芬兰混凝土建筑》，尤西·第艾宁拍摄，拉肯努斯帝托出版社，芬兰，赫尔辛基，2008 年，应急服务学院，第 46 ～ 53 页，参议院办公大楼，第 80 ～ 87 页；
《地球建筑》，罗纳德·雷亚尔著，普林斯顿建筑出版社，美国，纽约，2008 年，埃拉别墅，第 160 ～ 163 页，卡荷尔埃拉家禽养殖学校，第 164 ～ 167 页；
《芬兰设计——手工整形》，威林古斯出版社，2008 年；《尤哈·艾林宁：建筑与设计的前卫联盟》，海门林纳省档案馆，第 253 页；
《芬兰建筑》，埃亚·罗斯肯，芬兰建筑博物馆，芬兰，赫尔辛基，2008 年；赫尤里卡，芬兰驻美国华盛顿大使馆，第 126 页和第 146 页；
《建筑玫瑰》，赫尔辛基城市建筑控制委员会，芬兰，赫尔辛基，2008 年；瓦萨里球场入口设计，第 46 ～ 47 页，参议院办公大楼与国家福利与健康研究中心，第 88 ～ 91 页。

2007 年

《1000 个欧洲建筑》，沃拉格绍斯·布劳恩出版社，2007 年，参议院办公大楼与国家福利与健康研究中心，第 54 页；
《现代建筑 A ～ Z》，彼得·高赛尔编，塔森出版社，德国，科隆，2007 年；
《科米》，乔治·施泰因曼，斯塔姆菲出版社，瑞士，伯尔尼，2007 年，科米项目；
《砖上的文字》，Conarquitectura 出版，西班牙，马德里，2007 年，参议院办公大楼与国家福利与健康研究中心，第 96 ～ 104 页；
《芬兰办公大楼》，ATL 和建筑信息出版社，芬兰，赫尔辛基，2007 年；参议院办公大楼与国家福利与健康研究中心，第 16 ～ 29 页；
《设计与木材》，卢克·贝尔塔和马可·波瓦蒂，马吉奥利出版社，2007 年，桑拿西蒙斯，第 163 ～ 166 页；
《科学空间》，Daab 出版社，德国，科隆，2007 年；马克斯普朗克分子生物学和遗传学研究院，第 90 ～ 95 页；
《世界建筑师事务所——51 种设计概念与作品》，渊上正幸著，ADP 出版社，日本，东京，2007 年，第 70 ～ 75 页。

2006 年

《美国当代建筑指南（美国东部第 2 卷）》，渊上正幸，东陶出版社，日本，东京，2006 年；

2009

Cultural Centres, architecture 1990-2011, Cecilia Bione, Motta, Milan 2009, p104-109, Juminkeko-Information Centre for the Karelian Culture, Kuhmo
4×4. Architectural Crossroads, Marco D'Annuntiis and Sara Cipolletti Quodlibet Studio. Architettura Ascoli Piceno 2009
Tehdään elementeistä, Suomalaisen betonielementtirakentamisen lähihistoria, SBK-säätiö, Jyväskylä 2009; p182-183, Finnish Science Center

2008

Nordic Architects Write, edited by Michael Asgaard Andersen, Routledge, Abingdon 2008, p183-188, Markku Komonen: Construction, Technolgy and Art
Concrete Architecture in Finland Photographed by Jussi Tiainen, Rakennustieto Publishing, Helsinki 2008, p46-53, Emergency Sevices College; p80-87, Senate Properties Office Building
Earth Architecture, Ronald Real, Princeton architectural Press, New York 2008, p160-163, Villa Eila; p164-167, Kahere Eila Poultry Farming School
Suomalainen muotoilu- käsityöstä muotoiluun, Weilin+Göös 2008; Juha Ilonen: *Arkkitehtuurin ja muotoilun särmikäs liitto*, p253, Hämeenlinna Provincial Archive
Finnish Architecture, Eija Rauske, Museum of Finnish Architecture, Helsinki 2008; p126, 146, Heureka, Finnish Embassy in Washington, Juminkeko
The Rose for Building, Helsinki City Building Control Comission, Helsinki 2008; p46-47, Vuosaari Gateway, p88-91, Stakes and Senate Headquarters

2007

1000×European Architecture, Verlagshaus Braun, 2007; p54, Stakes and Senate Properties office buildings
Modern Architecture A-Z, edited by Peter Gössel, Taschen, Cologne 2007
Komi, George Steinmann, Stämpfli, Bern 2007; Komi project
Words on Brick, conarquitectura ediciones, Madrid 2007; p96-104, Stakes and Senate Properties office buildings
Office Buildings in Finland, ATL and Rakennustieto Publishing, Helsinki 2007; p16-29, Stakes and Senate Properties office buildings
Progettare con il legno, Luca Berta and Marco Bovati, Maggioli Editore, 2007; p163-166, Sauna Simons
Science Spaces, daab, Cologne 2007; p90-95, Max Planck Institute of Molecular Cell Biology and Genetics
World Architects- 51 Concepts and Works, Masayuki Fuchigami, ADP, Tokyo 2007; p70- 75

2006

A guide to contemporary Architecture in America, Volume 2 Eastern U.S.A., Masayuki Fuchigami, Toto, Tokyo 2006

赫尔辛基城市指南》，赫尔辛基城市规划部，芬兰，赫尔辛基，
06 年，乌萨里球场入口设计，第 158 页；
芬兰别墅与桑拿》，哈利・赫他维亚，拉肯努斯帝托出版社，
兰，赫尔辛基，2006 年；
钢铁视觉效果》，埃萨・皮隆内编，TEY 与艾维出版社，芬兰，
尔辛基，2006 年；
电影与建筑》，韩国釜山电影院建筑群国际设计邀请赛，2006 年；
GA 当代建筑 03 期——图书馆系列》，纪夫富川编辑与拍摄，
本，2006 年。

)05 年

我们的家园》，安雅・图米著，卡努斯塔罗有限公司出版，芬
，坦佩雷，2005 年；
模块化住宅最佳设计》，马丁・尼古拉斯・孔兹、米歇尔・加
亚多，AV 编辑出版社，德国，斯图加特，2005 年；
模块化住宅》，马丁・尼古拉斯・孔兹、米歇尔・加林多，
/ 编辑出版社，德国，斯图加特，2005 年；
芬兰建筑大纲》，尼塔・尼库拉，芬兰，奥塔瓦，2005 年；
库奥皮奥市建筑史》，海伦娜・列基，芬兰，库奥皮奥，2005 年；
设计手册: 研究与科技建筑》，哈登・布劳恩，迪特尔・格罗姆林，
克豪斯出版社，瑞士，巴塞尔，2005 年。

)04 年

芬兰的小房子》，埃萨・皮隆内，拉肯努斯帝托出版社，芬兰，
尔辛基，2004 年；
现代组合式预制房》，吉尔・赫伯斯，赫伯国际设计出版社，
国，纽约，2004 年；
当代建筑》，图像出版社，澳大利亚，马尔格雷夫，2004 年；
木制房屋》，费德里克莫塔编辑出版社，意大利，米兰，2004 年；
当代世界建筑菲登阿特拉斯集》，费顿出版社，英国，伦敦，
)04 年；
建筑中的金属外墙》，埃斯科・米耶蒂宁编，TRY 建筑信息出
社，芬兰，赫尔辛基，2004 年。

003 年

现代景观》，费顿出版社，英国，伦敦，2003 年；
当今世界住宅——当代建筑方向》，宇宙出版社，美国，纽约，
)03 年；
斯堪的纳维亚设计——家具与建筑》，珀蒂・格兰德出版社，
本，东京，2003 年；
当代建筑》，图像出版社，澳大利亚，马尔格雷夫，2003 年；
芬兰新建筑》，东南大学出版社，中国，南京，2002 年；
小房子》，尼古拉斯・波普尔，劳伦斯・金出版社，英国，伦
，2001 年。

002 年

世界当代住宅》，玛莎・托瑞丝・阿西拉编，阿卓姆出版社，
班牙，格拉纳达，2002 年；
当今建筑（第 12 卷）》，菲利普 ・赫迪多编，塔森出版社，德国，
隆，2002 年；

Urban Guide Helsinki, Helsinki City Planning Department, Helsinki 2006; p158, Vuosaari Gateway

Villas and Saunas in Finland, Harri Hautajärvi, Rakennustieto, Helsinki 2006

Steel Visions, Edited by Esa Piironen, TRY & Avain Publishers, Helsinki 2006

Film and Architecture, Busan Cinema Complex International Invited Competition, Korea 2006

GA Contemporary Architecture 03, Library, Edited and Photographed by Yukio Futagawa, Japan 2006

2005

Me teemme koteja, Anja Tuomi, Kannustalo Ltd., Tampere 2005

Best designed modular houses, Martin Nicholas Kunz, Michelle Galiado, avedition, Stuttgart 2005

Modular Houses, Martin Nicholas Kunz, Michelle Galindo, AV Edition, Germany 2005

Suomen arkkitehtuurin ääriviivat, Riitta Nikula, Otava 2005

Kuopion kaupungin rakennushistoria, Helena Riekki, Kuopion kaupunki, Kuopio 2005

A Design Manual: Research and Technology Buildings, Hardo Braun, Dieter Grömling, Birkhäuser, Basel 2005

2004

Small Houses in Finland, Esa Piironen, Rakennustieto, Helsinki 2004

Prefab modern, Jill Herbers, Harper Design International, New York 2004

CA-Contemporary Architecture, Images Publishing, Mulgrave, Australia 2004

Case in Legno, Federico Motta Editore, Milano 2004

The Phaidon Atlas of Contemporary World Architecture, Phaidon Press, London 2004

Metallijulkisivut arkkitehtuurissa, Edited by Esko Miettinen, TRY & Building Information, Helsinki 2004

2003

Modern Landscape, Phaidon, London 2003

World House Now - contemporary architectural directions, Universe Publishing, New York 2003

Scandinavian Design, Furniture & Architecture, Petit Grand Publishing, Tokyo 2003

Contemporary Architecture, The Images Publishing, Australia 2003

Finnish New Architecture, Southeast University press, China 2002

Small Houses, Nicolas Pople, Larence King, London 2003

2002

Contemporary Houses of the World, edited by Martha Torres Arcila, Atrium, Spain 2002

Architecture Now ! Volume 12, edited by Philip Jodido, Taschen, Köln 2002

《楼梯、设计和建造》，卡尔·J·赫伯曼，博克豪斯出版社，瑞士，巴塞尔，2002年；
《组合式预制房》，艾莉森·艾瑞弗和布赖恩·伯克哈特，吉布斯·史密斯出版社，美国，佛罗里达州雷顿，2002年；
《德国巴伐利亚州政府建筑2001》，巴伐利亚州内政部最高建筑管理局，德国，慕尼黑，2001年。

Staircases, Design and Construction, Karl J. Habermann, Birkhäus
Basel 2002
Prefab, Allison Arieff and Bryan Burkhart, Gibbs Smith, Layton 2002
Staatlicher Hochbau in Bayern 2001, Oberste Baubehörde im Bay
ischen Staatsministerium des Innern, München 2002

2001年

《XS——大思想，小建筑》，泰晤士和哈德森出版社，英国，伦敦，2001年；
《20世纪的建筑》，彼得·高赛尔、加布里埃莱·卢萨乌瑟，塔森出版社，德国，科隆，2001年；
《现代化与社区：伊斯兰世界的建筑》，肯尼斯·弗兰姆普顿、查尔斯·柯里亚、戴维·罗宾逊，泰晤士和哈德森出版社，英国，伦敦，2001年；
《建筑学》，考尼曼恩出版社，德国，科隆，2001年；
《钢铁》，芬兰建筑钢结构协会，建筑信息有限公司出版，芬兰，赫尔辛基，2001年。

2001

XS - Big Ideas, Small Buildings, Thames & Hudson, London 2001
Architecture in the Twentieth Century, Peter Gössel, Gabriele Leuthä
er, Taschen, Köln
Modernity and Community: Architecture in the Islamic World, Kenne
Frampton, Charles Correa, David Robson, Thames & Hudson, Lond
2001
Arkkitehtuuriatlas, Könemann, Köln 2001
Steel, Finnish Constructional Steelwork Association, Building Informati
Ltd, Helsinki 2001

2000年

《伟大的建筑师》，阿卓姆出版社，西班牙，格拉纳达，2000年。

2000

Great Architets, Atrium, Grabasa 2000

1999年

《创意工作室》，杰里米·迈尔森和菲利普·罗丝，劳伦斯·金出版社，英国，伦敦，1999年；
《愿景——现代芬兰设计》，安尼·斯滕罗斯主编，芬兰，凯乌鲁，1999年；
《建立科学》，马克斯普朗克分子生物学和遗传学研究院，博克豪斯出版社，瑞士，巴塞尔，1999年。

1999

The Creative Office, Jeremy Myerson and Philip Ross, Laurence Kir
London 1999
Visioita. Moderni suomalainen muotoilu, Anne Stenros (ed.), Keur
1999
Bauen für Wissenschaft, Institute der Max-Planck-Gesellschaft, Birkhäu
er, Basel 1999

1998年

《感官建筑》，克里斯汀·W. 汤姆森，帕莱斯特出版社，美国，纽约，1998年。

1998

Sensuous Architecture, Christian W. Thomsen, Prestel, New York 199

1997年

《极简主义建筑》，弗朗西斯科·阿森西奥·切沃，赫斯特国际图书出版社，美国，纽约，1997年。

1997

The Architecture of Minimalism, Francisco Asensio Cerver, Hear
Books International, New York 1997

1996年

《建设的意义》，德国，慕尼黑，1996年；
《创新建筑》，弗朗西斯科·阿森西奥·切沃，西班牙，巴塞罗那，1996年；
《20世纪的现代建筑》，威廉·J·R. 柯蒂斯，费顿出版社，英国，伦敦，1996年；
《新公共建筑》，迈尔森，劳伦斯·金出版社，英国，伦敦，1996年。

1996

Bauen für die Sinne, München 1996
Innovative Architecture, Francisco Asensio Cerver, Barcelona 1996
Modern Architecture since 1900, William J.R. Curtis, Phaidon, Lond
1996
New Public Architecture, Myerson, Laurence King, London 1996

1995年

《恩格尔的遗产》，拉肯努斯帝托出版社，芬兰，赫尔辛基，1995年；
《芬兰建筑与现代主义传统》，马尔科姆·匡斯瑞尔，E&FN出版社，英国，伦敦，1995年；

1995

Engelin perintö, Rakennustieto Oy, Helsinki 1995
Finnish Architecture and the Modernist Tradition, Malcolm Quantri
E & FN, London 1995
Contemporary European Architects Volume IV, Taschen, Köln 1996, I
troduction by Philip Jodido

当代欧洲建筑师（第四卷）》，菲利普·赫迪多作序，塔森出社，德国，科隆，1996 年；
新世纪博物馆》，何塞普·M·蒙塔内，古斯塔沃·吉利出版社，班牙，巴塞罗那，1995 年；
581 位世界建筑师》，东陶舒普恩出版社，日本，东京，1995 年。

Museums for the New Century, Josep M. Montaner, Gustavo Gili, Barcelona 1995
581 Architects in the World, TOTO Shuppen, Tokyo 1995

94 年

困惑的建筑》，埃里克·尼加德，克里斯汀·伊勒斯出版社，威，奥斯陆，1994 年。

1994

Arkitektur i en forvirret tid, Erik Nygaard, Christian Ejlers, Oslo 1994

93 年

欧洲城市品质》，博洛尼亚国际银行，劳拉·佩德罗蒂，意大，博洛尼亚，1993 年。

1993

Qualità Urbana in Europa, Fiere Internazionali di Bologna, Laura Pedrotti, Bologna 1993

92 年

现代建筑——一部批判的历史》，肯尼斯·弗兰姆普顿，泰晤和哈德森出版社，英国，伦敦，1992 年；
新芬兰建筑》，斯科特·普尔，里佐利出版社，美国，纽约，92 年。

1992

Modern Architecture - A Critical History, Kenneth Frampton, Thames & Hudson, London 1992
The New Finnish Architecture, Scott Poole, Rizzoli, New York 1992

91 年

现代建筑》，马滕·克劳斯编，建筑和自然出版社，荷兰，阿斯特丹，1991 年；
欧洲大师 3》，阿卓姆出版社，西班牙，巴塞罗那，1991 年；
当代建筑 1990—1991 年》，技术与学术出版社，瑞士，洛桑，91 年；
建筑年鉴 1：欧洲大师》，阿卓姆出版社，西班牙，巴塞罗那，87 年。

1991

Architecture Now, Maarten Kloos (ed.) Architectura & Natura, Amsterdam 1991
European Masters/3, Atrium, Barcelona 1991
Contemporary Architecture 90-91, Presses Polytechniques et Universitaires Romandes, Lausanne 1991
European Masters, Annual of Architecture/1, Atrium, Barcelona 1987

展览会与研讨会

016 年

“北欧国家面对面疗愈”，威尼斯建筑双年展北欧展馆，2016 年，门林纳省档案馆和萨文林纳主图书馆。

015 年

“做！”，第 13 届国际阿尔瓦·阿尔托研讨会，第 46 ~ 51 页，“本地的吗？”，米高·海基宁。

014 年

12 届多科莫莫国际现代建筑文献会议记录，第 288 ~ 294 页，“老沟—新水”，米高·海基宁，多科莫莫，2014 年。

012 年

“2010—2011 年的芬兰建筑”，芬兰建筑博物馆等，2012 年，洛拉诺基奥住宅区设计，第 32 ~ 40 页。

Exhibition and symposium catalogues

2016

In Therapy: Nordic Countries Face to Face, The Nordic Pavilion exhibition at the Venice Architecture Biennale 2016, Hämeenlinna provincial archive and Savonlinna main library

2015

DO!, 13th International Alvar aalto Symposium, p46-51, “Local?” by Mikko Heikkinen

2014

Proceedings of the 12th International Docomomo Conference, “p288-294, “Old Ditch-New Water” by Mikko Heikkinen, do co mo mo 2014

2012

Finnish Architecture 2010/2011, Museum of Finnish Architecture et al 2012, Flooranaukio Housing p32-40

2010年

"芬兰建筑2008—2009"，阿尔瓦阿尔托学院，芬兰建筑师协会，芬兰建筑博物馆，芬兰，赫尔辛基，2006年。

2009年

《芬兰新兴建筑师》，第6～9页，"四十岁以下"，米高·海基宁作序；
《4×4建筑十字路口》，夸特罗建筑国际会议，阿斯科利皮切诺建筑公司，2009年。

2008年

《在那里——参与国家间独特的各项活动》，威尼斯双年展，意大利，马西利奥，2008年，沃塔洛文化中心；
《送货上门——装配现代住宅》，现代艺术博物馆，美国，纽约，2008年，科斯克特斯预制房屋；
《储存与共享——芬兰博物馆和图书馆》，芬兰建筑博物馆，芬兰，赫尔辛基，2008年，沃塔洛文化中心。

2007年

《芬兰建筑——教育和文化领域的当前项目》，ATL，芬兰建筑师协会（SAFA），德国，柏林，2007年。

2006年

《芬兰建筑2004—2005》，阿尔瓦阿尔托学院，芬兰建筑师协会，芬兰建筑博物馆，芬兰，赫尔辛基，2006年。

2005年

《木材建筑——从木材到建筑》，芬兰建筑博物馆，芬兰，赫尔辛基，2005年。

2004年

《蜕变——第九届国际建筑展》，威尼斯建筑双年展，意大利，威尼斯，2004年；
《芬兰建筑2002—2003》，阿尔瓦阿尔托学院，芬兰建筑师协会，芬兰建筑博物馆，芬兰，赫尔辛基，2004年；
《大象与蝴蝶——建筑的持久与机遇》，第九届阿尔瓦·阿尔托研讨会，米高·海基宁编，阿尔瓦阿尔托学院，芬兰，赫尔辛基，2004年；
《赫尔辛基当代城市建筑》，尤西·第艾宁拍摄，建筑信息，芬兰，赫尔辛基，2004年。

2003年

《共鸣——跨越地平线》，弗雷姆出版社，芬兰，图尔库，2003年
《雪中的威尼斯》，活力杂志，美国，纽约，2003年；
《Q——设计宁静——当代芬兰设计》，日本—芬兰设计协会，芬兰，赫尔辛基，2003年。

2010

Finnish Architecture 0809, Alvar Aalto Academy, Finnish Association Architect, Museum of Finnish Architecture, Helsinki 2006

2009

Newly Drawn-Emerging Finnish Architects, p6-9, Introduction "Und Forty" by Mikko Heikkinen
4×4 Architectural Crossroads, Quattro conference internazionali di a chitettura, Architettura Ascoli Piceno 2009

2008

Out There- Participating Countries Special and Collateral Events, l Biennale di Venezia, Marsilio 2008, Vuotalo Cultural Center
Home Delivery- Fabricating Modern Dwelling, The Museum of Mode Art, New York 2008, House Kosketus
Store and Share- museums and libraries in Finland, Museum of Finnis Architecture, Helsinki 2008, Vuotalo Cultural Center

2007

Finnische Architektur- Aktuelle Projekte im Bildungs- und Kulturber ich, ATL, SAFA, Berlin 2007

2006

Finnish Architecture 0405, Alvar Aalto Academy, Finnish Association Architect, Museum of Finnish Architecture, Helsinki 2006

2005

Arkkitehtuuria puusta - From wood to architecture, Museum of Finnis architecture, Helsinki 2005

2004

Metamorph - 9th International Architecture Exhibition, La Biennale Venezia, Venice 2004
Finnish Architecture 0203, Alvar Aalto Academy, Finnish Association Architects, Museum of Finnish Architecture, Helsinki 2004
Elephant & Butterfly, Permanence and Chance in Architecture - 9th Alva Aalto Symposium, edited by Mikko Heikkinen, Alvar Aalto Academy, Helsinl 2004
Helsinki Contemporary Urban Architecture photographed by Jussi T ainen, Building Information, Helsinki 2004

2003

Empathy - Beyond the Horizon, Frame Publications, Turku 2003
The Snow Show Venice, Zingmagazine Books, New York 2003
Q - Designing the Quietness - Contemporary Finnish Design, Japan Fin land Design Association, Helsinki 2003

002 年

当代芬兰建筑》，尤西・第艾宁拍摄，建筑信息，芬兰，坦佩，2002 年；
下一个——第八届国际建筑展”，威尼斯双年展，意大利，威斯马西利奥，2002 年；
下一步之前——从基础学起”，芬兰建筑博物馆，芬兰，赫尔基，2002 年。

001 年

赫尔辛基：当代城市建筑》，尤西・第艾宁拍摄，建筑信息，兰，赫尔辛基，2004 年；
格式部》，尤西・第艾宁拍摄，西班牙，马德里，2001 年，
5×5 50 年的照片》，米高・海基宁。

000 年

00 年国际舞美组织出版影院，德国，慕尼黑，2000 年；
德巴萨贝尔国际挑战赛，雷克雅未克美术馆，冰岛，雷克雅未，2000 年；
20 世纪芬兰建筑”，芬兰建筑博物馆，芬兰，赫尔辛基，德建筑博物馆，德国，法兰克福，2000 年。

999 年

芬兰建筑 1994—1999”，尤西・第艾宁拍摄，芬兰建筑中心，99 年。

998 年

芬兰建筑 9”，芬兰建筑博物馆，芬兰，于韦斯屈莱，1998 年。

996 年

斯・凡・德罗欧洲建筑展馆奖，西班牙，巴塞罗那，1996 年。
考林尼哈文国际挑战赛》，罗德斯国际科学和艺术出版社，丹，哥本哈根，1996 年；
芬兰木制建筑”，芬兰建筑博物馆，芬兰，赫尔辛基，1996 年；
新芬兰建筑”，尤西・第艾宁拍摄，芬兰建筑中心，芬兰，赫辛基，1996 年。

995 年

为建筑而设计”，阿尔瓦阿尔托博物馆，芬兰，于韦斯屈莱，995 年。

992 年

五届国际阿尔瓦・阿尔托研讨会，芬兰，于韦斯屈莱，1992 年；
对比与联系”，1992 年芬兰世博会，芬兰，赫尔辛基，1992 年；
斯・凡・德罗欧洲建筑展馆奖，西班牙，巴塞罗那，1992 年；
芬兰建筑 8”，芬兰建筑博物馆，芬兰，赫尔辛基，1992 年；
马德里的视觉故事”，五个建筑理念，西班牙，马德里，1992 年。

2002

Contemporary Finnish Architecture Photographed by Jussi Tiainen, Building Information, Tampere 2002
Next - 8th International Architecture Exhibition, La Biennale di Venezia, Marsilio, Venice 2002
Before Next - Learning from Roots, Museum of Finnish Architecture, Helsinki 2002

2001

Helsinki, Contemporary Urban Architecture photographed by Jussi Tiainen, Building Information, Helsinki 2001
Jussi Tiainen Fotografias, Ministerio de Formato, Madrid 2001, essay by Mikko Heikkinen: "Cinco por Cinco - fotografias de cinco décadas"

2000

OISTAT Publication Theatres in 2000, Munich 2000
Gardhúsabaer, The International Challenge, Reykjavik Art Museum, Reykjavik 2000
20th Century Architecture Finland, Museum of Finnish Architecture, Helsinki, Deutsches Architektur-Museum, Frankfurt am Main, 2000

1999

Finnish Architecture 1994-1999, photographed by Jussi Tiainen, Finnish Building Center 1999

1998

Finland Builds 9, Museum of Finnish Architecture, Jyväskylä 1998

1996

Mies van der Rohe Pavilion Award for European Architecture, Barcelona 1996
Kolonihaven, The International Challenge, Rhodos International Science and Art Publishers, Copenhagen 1996
Timber Construction in Finland, Museum of Finnish Architecture, Helsinki 1996
New Finnish Architecture Photographed by Jussi Tiainen, Finnish Building Center, Helsinki 1996

1995

Design for Architecture, Alvar Aalto Museum, Jyväskylä 1995

1992

The 5th International Alvar Aalto Symposium, Jyväskylä 1992
Contrasts & Connections, Expo '92 Finlandia, Helsinki 1992
Mies van der Rohe Pavilion Award for European Architecture, Barcelona 1992
Finland Builds 8, Museum of Finnish Architecture, Helsinki 1992
Visiones para Madrid, Cinco ideas Arquitectonicas, Madrid 1992

1991年

1991年布拉格研讨会，捷克，布拉格，1991年。

1990年

“11座城市，11个国家，当代北欧艺术和建筑”，荷兰，吕伐登，1990年；
“第四纪1990”，国际建筑创新技术奖，意大利，威尼斯，1990年；
“现代芬兰”，建筑博物馆，瑞典，斯德哥尔摩，1990年；
“1900—1990芬兰建筑、手工艺品和绘画”，德国，慕尼黑，1990年；
“建筑现状——七种方法”，芬兰建筑博物馆，芬兰，赫尔辛基，1990年。

1989年

“建筑、工艺和设计”，建筑、城市规划和设计国际研讨会，芬兰，艾斯堡，1989年；
“芬兰建筑混凝土”，芬兰混凝土工业协会，芬兰建筑博物馆，芬兰，赫尔辛基，1989年。

1991

Workshop Prague ‘91, Prague 1991

1990

11 Cities, 11 Nations, Contemporary Nordic Art and Architecture, Leeuwarden 1990
Quaternario 90, International Award for Innovative Technology in Architecture, Venice 1990
Finskt I Nuet, Arkitektur Museet, Stockholm 1990
Architektur, Kunsthandwerk, Malerei, Finnland 1900-1990, Munich 1990
An Architectural Present - 7 Approaches, Museum of Finnish Architecture, Helsinki 1990

1989

Architecture, Craftmanship and Design, International Conference on Architecture, Urban Planning and Design, Espoo 1989
Concrete in Finnish Architecture, Association of the Concrete Industries of Finland, Museum of Finnish Architecture, Helsinki 1989

建筑杂志和网络出版物的报道

2016年

《DIVISING》，芬兰康阿斯阿拉艺术中心；
《北欧国家面对面疗愈》，威尼斯建筑双年展北欧馆展览，海门林纳省档案馆和萨文林纳主馆，2016年；
《威尼斯2016年双年展AIAC——A10出版物》，芬兰康阿斯阿拉艺术中心。

2015年

《建筑师》2015年第6期，第24～31页，斯托布尔公园，珍妮·肯塔拉编；
《建筑师》2015年第4期，第66～74页，芬兰康阿斯阿拉艺术中心和基莫·皮耶克艺术博物馆，奥利—帕沃·科波宁编。

2014年

《建筑师》2014年第3期，第46～54页，新萨文林纳主图书馆，《从白鲑鱼到冻鱼条》，尤西·沃里编。

2013年

《建筑实录》2013年第03期，第106～108页，“赫尔辛基弗洛拉诺基奥住宅区设计的地方特色”，彭蒂·科瑞奥亚编；
《建筑师》2013年第03期，第56～62页，木材表面，罗伊·曼塔里，弗里达别墅；
《Domus》第972期，第1～8页，“面向竞赛的建筑”，肯尼斯·弗兰姆普顿，非洲项目。

Architectural Magazines and web publications

2016

DIVISARE (Europaconcorsi), Kangasala Arts Centre
In Therapy: Nordic Countries Face to Face, The Nordic Pavilion exhibition at the Venice Architecture Biennale 2016, Hämeenlinna provincial archive and Savonlinna main library
AIAC-A10 publication for 2016 VENICE BIENNALE, Kangasala Arts Centre

2015

Arkkitehti 6/2015, p24-31,Schönbühl Park, text by Janne Kentala
Arkkitehti 4/2015, p66-74, Kangasala Arts Centre and Kimmo Pyykkö Art Museum, text by Olli-Paavo Koponen

2014

Arkkitehti 3/2014, p46-54, New main library of Savonlinna, “From vendance to fishfinger” by Jussi Vuori

2013

Architectural Record 03/2013, p106-108, “Local Identity,Helsinki Flooranaukio Housing” by Pentti Kareoja
Arkkitehti 3/2013, p56-62, “Pintaa puusta” by Roy Mänttäri, Villa Frida
Domus 972, p1-8,”Towards an Agonistic Architecture” by Kenneth Frampton, African projects

012 年

木质建筑》2012 年第 4 期，第 14 ～ 17 页，“接触住房”；
建筑师》2012 年第 3 期，第 54 ～ 60 页，弗洛拉诺基奥住宅
设计；
形式》2012 年第 1 期，第 088 ～ 095 页，弗洛拉诺基奥住宅
设计；
426 建筑业，ANCE》，第 112 ～ 116 页，“芬兰建筑 2010—
11”，弗洛拉诺基奥住宅区设计，瑞玛托·莫尔甘蒂和达尼
·迪·多纳托撰写；
装饰》2012 年第 8 期，第 42 ～ 48 页，洛赫涅米别墅。

011 年

建筑师》2011 年第 6 期，第 11 ～ 21 页，“运动和反运动：芬
后现代主义印象”，安妮·瓦托拉撰写；
世界建筑》主题 11：博物馆大厦，第 18 页，曼塔格拉斯塔斯
物馆，第 34 页，努克国家美术馆；
世界建筑》2011 年第 8 期，第 33 页，格拉斯塔斯博物馆，曼塔；
建筑师》2011 年第 3 期，第 22 ～ 25 页，“驯鹿的粪便——
建筑的随评”，米高·海基宁；
世界建筑》杂志 2011 年第 4 期，第 36 页，努克国家美术馆；
小木屋与木房子》，2011 年 2 月～ 3 月第 49 期，第 64 ～ 65 页，
触住房。

010 年

德国建筑报》2010 年第 12 期，第 12 ～ 13 页，“更多的光”，
尔马·姆卡拉，海门林纳省档案馆；
建筑师》2010 年第 6 期，“纸老虎”，第 42 ～ 43 页，釜山
影院建筑群；
C3》2010 年 10 月第 314 期，第 94 ～ 99 页，海门林纳省档案馆；
建筑评论》2010 年第 5 期，第 68 ～ 73 页，海门林纳省档案馆；
建筑实录》2010 年第 02 期，第 43 页，海门林纳省档案馆，“让
泥板焕然一新”，丽塔·卡廷拉·奥瑞尔；
建筑师》2010 年第 1 期，第 24 ～ 38 页，海门林纳省档案馆。

009 年

《论坛》2009 年第 3 期，第 102 ～ 113 页，芬兰驻华盛顿大使馆，
“有着最多大使馆的国家首都”，拉尔斯·福斯贝里；
《建筑 DK》2009 年第 6 期，第 A2 ～ A10 页，海门林纳省档案馆，
“沉重的图形”，麦克·罗密尔；
《A10》2009 年 5 月 / 6 月，第 64 ～ 66 页，弗洛拉诺基奥住宅
设计，“三座城市的故事”，塔里亚·努尔米；
《未来》19/20，第 126 ～ 127 页，芬兰驻东京大使馆，参赛作品
“汉卡”；
《木质建筑》2009 年第 4 期，第 44 ～ 45 页，考帕民乐学院，芬兰，
赫莫；
《实验室 09》第 7 期，第 4 页，“定位神圣——访问马尔库·科
宁”，凯蒂·布鲁卡撰写；
《实验室 09》第 6 期，第 2 页，“弗雷德建筑：马尔库·科莫
对话佩佩·巴尔贝里”，古斯佩·拜罗纳多撰写；

2012

PuuWoodHolzBois 4/2012, p14-17, House Touch
Arkkitehti 3/2012, p54-60, Flooranaukio Housing
Form 1/2012, p88-95, Flooranaukio Housing
426 l'industria delle costruzioni, ANCE, p112- 116 “Architettura finlandese 2010-2011” by Remato Morganti and Danilo di Donato, Flooranaukio Housing
deko 8/12, p42-48, Villa Louhenniemi

2011

Arkkitehti 6/2011, p11-21, “Movement and counter-movement: Impressions of Finnish potsmodernism” by Anni Vartola
wa Themenbuch 11 – Museumsbauten, p18, Serlachius Museum Gösta in Mänttä, p34, National Gallery of Art in Nuuk
wa 8/2011, p33, Serlachius Museum Gösta in Mänttä
Arkkitehti 3/2011, p22-25, “Reindeer Droppings – Random Remarks on Architecture” by Mikko Heikkinen
wa 4/2011, p36, National Gallery of Art in Nuuk
Chalets & Maisons BOIS, 49/Février-Mars 2011, p64-65, House Touch

2010

db /Deutsche Bauzeitung 12/2010, p12-13, “Mehr Licht” by Jorma Mukala, Hämeenlinna Provincial Archive
Arkkitehti 6/2010, p42-43, “Paper Tigers”, Busan Cinema Complex
C3, 314/Oct 2010, p94-99, Hämeenlinna Provincial Archive
Architectural Review 5/2010, p68-73, Hämeenlinna Provincial Archive
Architectural Record 02/2010, p43, Hämeenlinna Provincial Archive, “Putting a fresh face on concrete panels” by Rita Catinella Orrell
Arkkitehti 1/2010, p24-38, , Hämeenlinna Provincial Archive

2009

Forum 3/2009, p102-113, Finnish Embassy in Washington, “Ambassaarinen, eller Varför flyttar allt fler länder ut från huvudstäderna” by Lars Forsberg
Arkitektur DK 6/09, pA2-A10, Hämeenlinna Provincial Archive, “Tung grafik” by Mike Römer
A10 May/June 2009, p64-66, Flooranaukio Housing Block, “A Tale of Three Cities” by Tarja Nurmi
Future 19/20, p126-127, Embassy of Finland in Tokyo, Competition entry “Hanka”
Puu-Wood-Holz-Bois 4/2009, p44-45, Folk Music Academy Koppa, Kuhmo
Laboratorio 09, numero 7, p4, “Locating Sacredness, interview with Markku Komonen” by Ketty Brocca
Laboratorio 09, numero 6, p2, “Architetture fredde, Markku Komonen vs Pepe Barbieri” by Giuseppe Peronato

《实验室 09》第 0 期，第 5 页，“神圣空间”研讨会，马尔库·科莫宁；
《建筑师》2009 年第 3 期，第 11 ~ 17 页，海门林纳省档案馆，“装饰品返回到犯罪现场”，约尼·鲁斯。

2008 年

《绿色资源》2008 年 9 月，芬兰驻华盛顿大使馆，“翠绿的表面”，B·J·诺维茨基；
《埃希特快捷号》2008 年 4 月，第 8 ~ 10 页，拉乌预制房屋，“教师的最爱：紧凑的拉乌模型”，阿格妮塔·兰德；
《建筑师》2008 年第 5 期，第 44 ~ 45 页，蒙纳提住宅；
《建筑师》2008 年第 4 期，第 79 页，赫尔辛基大学中心校区图书馆大赛参赛作品，“偷工减料”，克里斯托·瓦斯卡沙；
《斯卡拉马拉》2008 年第 09 期，第 33 页，科斯克特斯预制房屋；
《建筑评论》2008 年第 5 期，第 54 页，“问题评论：海门林纳省档案馆”；
《阿卓姆》2008 年第 01 期，第 62 ~ 67 页，桑拿西蒙斯，“一个新的模型”，佩托·包德沃。

2007 年

《建筑师》2007 年第 5 期，第 80 页，拉乌预制房屋，“精美的原始小屋”，克里斯托·瓦斯卡沙；
《建筑师》2007 年第 2 期，第 70 ~ 71 页，阿拉比亚沿海房屋“洛伦斯”设计竞赛，“房屋结构的变形问题”，约尔马·姆卡拉；
《建筑业》396 期，第 90 页，拉彭兰塔理工大学第 7 期，“芬兰建筑 2004—2005”，雷纳托·莫尔甘蒂；
《PUU》2007 年第 3 期，第 32 ~ 33 页，“拉乌预制房屋”；
《建筑师》2007 年第 11 期，第 54 ~ 56 页，马克斯普朗克分子生物学和遗传学研究院，“当前医疗建筑的设计趋势”，卡雷尔·弗特尔；
《论坛辅刊》2007 年第 4 期，第 84 ~ 91 页，拉乌预制房屋，“阿尔瓦·阿尔托的遗产”，约翰娜科尔乔宁；
《OZ 卷》2007 年第 29 期，第 36 ~ 41 页，“空间、时间和建筑”，米高·海基宁；
《建筑评论》2007 年第 1 期，第 70 页，“为海门林纳房屋展览会建造的预制房屋”。

2006 年

《iA 室内设计 / 建筑》2006 年第 2 期，第 123 ~ 128 页，参议院办公大楼与国家福利与健康研究中心，《芬兰》，麦古米·奥库伯；
《论坛》2006 年第 2 期，第 152 ~ 153 页，应急服务学院第 4 期；
《未来建筑》2006 年第 2 期，第 68 ~ 81 页，釜山复合式电影院，韩国首尔；
《空间》2006 年第 1 期，第 132 ~ 173 页，“权力采访：在韩国的外国建筑师”，林真英主编；
《建筑师》2006 年第 1 期，第 58 ~ 63 页，应急服务学院第 4 期，塔里亚·努尔米评论。

Laboratorio 09, numero 0, p5, “Sacred Space” Workshop by Markku K
monen
Arkkitehti 3/2009, p11-17, Hämeenlinna Provincial Archive, ”Orname
Returns to the Scene of Crime” by Jonni Roos

2008

Green Source September 2008, Finnish Embassy in Washington, “Ve
dant Surfaces” by B.J. Novitski
Ehitus Ekspress April 2008, p8-10, Laavu prefab home, “Tegijate le
mik: kompaktne moodulmaja Laavu” by Agneta Land
Arkkitehti 5/2008, p44-45, Sonaatti House
Arkkitehti 4/2008, p79, entry for the Helsinki University Central Ca
pus Library competition “Cutting corners” by Kristo Vesikansa,
Scanorama 9/08, p33, House Kosketus
Architectural Review 5/2008, p54, Preview Issue: Hämeenlinna Provi
cial Archive
Atrium 01/2008, p62-67, Sauna Simons, “Nový Archetyp” by Petra Bo
dová

2007

Arkkitehti 5/2007, p80, Laavu prefab home, “Refined Primitive Hut” b
Kristo Vesikansa
Arkkitehti 2/2007, p70-71, Arabianranta Housing Competition “Lo
rens”, “A Metamorphosis in housing architecture” by Jorma Mukala
L’industria delle costruzioni 396, p90, Lappeenranta University of Tec
nology, phase 7, “Architettura finlandese 2004-2005” by Renato Morgan
PUU 3/2007, p32-33, Laavu prefab home
Architekt 11/2007, p54-56, Max Planck Institute of Molecular Cell Bio
ogy and Genetics, “Soucasné tendence v architektoniké tvorbe zdravo
nických staveb” by Karel Fortl
Forum Aid 4.07, p84-91, Laavu prefab home, “Alvar Aalto’s Legacy” b
Johanna Koljonen
OZ Volume 29/2007, p36-41, “Space, Time and Architecture” by Mikk
Heikkinen
Architectural Review 1/2007, p70, Pre-fab House for the Hämeenlinn
Housing Fair

2006

iA interior/Architecture 2/2006, p123-128, Stakes and Senate Propertie
office buildings, “Finland” by Megumi Okubo
Forum 2/2006, p152-153, Emergency Services College Phase 4.
Future Arquitecturas 2/2006, p68-81, Busan Cinema Complex, Seou
Korea
Space 1/2006, p132-173, Power interview: foreign architects in Kore
edited by Lim Jinyoung
Arkkitehti 1/2006, p58-63, Emergency Services College Phase 4, criti
by Tarja Nurmi

005 年

平面设计》2005 年第 009 期，第 43 ～ 55 页，考利则基竞赛；
当前竞赛》2005 年第 3 期，第 43 ～ 52 页，考利则基竞赛；
md》2005 年第 8 期，第 38 ～ 41 页，拉彭兰塔理工大学第 7 期，
“北方的凉爽”，乌尔瑞其・布特奈尔；
大都市》2005 年第 4 期，第 106 ～ 107 页，瓦萨里球场入口设计，
“明亮的照明和硕大的城市”，克里斯迪・卡梅伦；
奥利斯（Oris）32》，第 88～101 页，“回到未来”，佩托・赛弗瑞；

建筑比赛》2005 年第 3 期，第 30 ～ 33 页，马克斯普朗克分
生物学和遗传学研究院；
卡萨・布鲁特斯》2005 年第 59 期，第 91 页，科斯克特斯预
房屋；
论坛》2005 年第 1 期，第 47 页，拉彭兰塔理工大学第 7 期，
萨・拉克索宁评论；
房屋》2005 年第 5 期，第 124 ～ 127 页，参议院办公大楼与
家福利与健康研究中心，“另一个机会 - 城市中心废弃工业园
的新生活”，玛特乌兹・泽里克；
建筑师》2005 年第 1 期，第 74 ～ 76 页，考利则基竞赛，“成
的痛苦”，佩托・赛弗瑞。

004 年

奥利斯》第 30 期，第 94 ～ 111 页，“考利则基竞赛”，鲍里
・波德莱卡；
《SD》2004 年，第 093 页，埃拉别墅；
建筑师》 2004 年第 6 期，第 62 ～ 67 页，拉彭兰塔理工大学
7 期；
今日建筑》第 153 期，2004 年第 10 期，第 60 ～ 64 页，拉彭
塔理工大学第 7 期；
建筑师》2004 年第 2 期，第 34 ～ 37 页，马克斯普朗克分子
物学和遗传学研究院；
建筑师》2004 年第 1 期，第 66 ～ 69 页，设计宁静的展示空间，
光疗法》，哈利・赫他维亚；
《心灵、土地与社会》，2004 年 7 月，第 59 ～ 62 页，第
08 ～ 111 页，“艺术与建筑之间”，米高・海基宁；
建筑业》2004 年第 376 期，第 90 ～ 96 页，几内亚项目，“非
的五个项目”，雷纳托・莫尔甘蒂；
巴西卡萨时尚》2004 年第 226 期，第 172 ～ 175 页，埃拉别墅；
“建筑——心灵、土地与社会”，2004 年 2 月，第 15 ～ 27 页，
瓦涅米机场航站楼和几内亚项目，《语境的重要性》，路易斯・安
尔・多明格斯；
小屋》2004 年第二期第一册，第 28 ～ 31 页，参议院办公大
与国家福利与健康研究中心；
居住在地球上》2004，第 30 ～ 33 页，几内亚项目，“非洲建
”，恩里科・斯西格纳诺。

003 年

项目》2003 年第 15 期，第 4 ～ 11 页，米高・海基宁和马尔库・科
宁，派塔・西凡瑞采访整理；

2005

av projectos 009/2005, p43-55, Kolizej competition
Wettbewerbe aktuell 3/2005, p43-52, Kolizej competition
md 8/2005, p38-41, Lappeenranta University of Technology, phase 7, "Northern Cool" by Ulrich Büttner
Metropolis 4/2005, p106-107, Vuosaari Gateway, "Bright Lights, Big City" by Kristi Cameron
Oris 32, p88-101, "Back to the Future" by Petra Ceferin

Architektur Wettbewerbe 3/2005, p30-33, Max Planck Institute of Molecular Cell Biology and Genetics
Casa Brutus 59/2005, p91, House Kosketus
Forum 1/2005, p47, Lappeenranta University of Technology, phase 7, critic by Esa Laaksonen
Hiše 5/2005, p124-127, Senate Properties office Building, "Druga priložnost - Novo življenje za opuščene industrijske komplekse sredi mesta" by Matevž Čelik
Arkkitehti 1/2005, p74-76, Kolizej competition, "Growing pains" by Petra Ceferin

2004

Oris 30, p94-111, "Kolizej area competition" by Boris Podrecca
SD 2004, p093, Villa Eila
Arkkitehti 6/2004, p62-67, Lappeenranta University of Technology, phase 7
Architecture Today 153, 10/2004, p60-64, Lappeenranta University of Technology, phase 7
Arkkitehti 2/2004, p34-37, Guesthouses of Max Planck Institute of Molecular Cell Biology and Genetics
Arkkitehti 1/2004, p66-69, Designing the Quietness Exhibition, "Light therapy" by Harri Hautajärvi
Mind, Land & Society, July 2004, p59-62, p108-111, "Between Art and Architecture" by Mikko Heikkinen
L'industria delle costruzioni 376/2004, p90-96, Guinean projects, "Cinque progetti in Africa" by Renato Morganti
Casa Vogue Brasil 226/2004, p172-175, Villa Eila
Arquitectonics - Mind, Land & Society, February 2004, p15-27, Rovaniemi Airport Terminal and Guinean projects, "De la necesidad del contexto" by Luis Angel Dominguez
Maja 1/2-2004, p28-31, Stakes and Senate Properties office buildings
Abitare la Terra 2004, p30-33, Guinean projects, "Architectures in Africa" by Enrico Sicignano

2003

il Progetto 15/2003, p4-11, Mikko Heikkinen and Markku Komonen, interview by Petra Čeferin

《日本建筑师》2003年第12期，第88～89页，"设计宁静"展览；
《建筑》2003年第8期，第62～67页，参议院办公大楼与国家福利与健康研究中心，"与谷物同行"，凯茜·朗·霍；
《AVS》2003年第3期，第46～47页，几内亚项目，"由现在构成的建筑作品"，哈多拉·阿纳多特；
《自然》第424期，第718～720页，"你想在这里工作吗？"，劳拉·波奈塔，第858～859页，"大胆的建筑旨在……"，肯德尔·鲍威尔，这两篇文章都提到了马克斯普朗克分子生物学和遗传学研究院；
《芬兰形式功能》，第6～9页，参议院办公大楼与国家福利与健康研究中心；
《雅皮》第261期，第66～70页，卢米媒体中心；
《方舟》第182期，第54～63页，马克斯普朗克分子生物学和遗传学研究院，"自然和技术之间"，伊莎贝尔·戈尔德曼；
《建筑业》第370期，第32～41页，卢米媒体中心，"赫尔辛基艺术与设计大学的多媒体教育中心"，雷纳托·莫尔甘蒂；
《莲花》第116期，第60～71页，几内亚项目，"建筑师和人类学家"，卡罗利那·弗依斯；
《雅皮》第258期，第80～87页，马克斯普朗克分子生物学和遗传学研究院；
《GA房屋》第74期，"2003年项目"，第20～21页，哈维纳别墅；
《AD第73卷》2003年第1期，"离开雷达"，布莱恩·卡特和安妮特·勒奎尔编，第26～28页，卡荷尔埃拉家禽养殖学校；
《建筑实录》2003年第1期，第110～117页，马克斯普朗克分子生物学和遗传学研究院，"科学家的社区意识"，戴维·科恩；
《建筑》2003年第1期，第100页，马克斯普朗克分子生物学和遗传学研究院，"楼梯"，芭芭拉·莱斯尼克。

2002年

《建筑师》2002年第6期，第28～37页，参议院办公大楼与国家福利与健康研究中心；
《建筑》2002年第11期，第76～79页，科斯克特斯预制房屋，"轻触"，C.C.·苏利文；
《钢铁方面的创新2002/教育设施》，第21页，卢米媒体中心；
《当前竞赛》2002年第1期，第57页，巴伐利亚州议会大厅参赛作品，慕尼黑；
《建筑评论》2002年第6期，第40～45页，沃塔洛文化中心，《很酷的都市风格》，亨利·迈尔斯；
《建筑评论》2002年第6期，第52～55页，马克斯普朗克分子生物学和遗传学研究院，"功能生物学"，蕾拉·道森；
《md》2002年第8期，第27～31页，卢米媒体中心，"阿拉伯海岸的新建筑"，苏珊娜·塔姆波瑞尼；
《阿尔瓦阿尔托公司协会第15公告/过梁》2002年，马克斯普朗克分子生物学和遗传学研究院；
《雅皮》2002年第8期，第61～68页，沃塔洛文化中心；
《建筑及论坛》，2002年第3～4期，第61页，尤米凯科；
《建筑师》2002年第2期，第32～41页，马克斯普朗克分子生物学和遗传学研究院；

Japan Architect 12/2003, p88-89, Designing the Quietness - Exhibition
Architecture 8/2003, p62-67, Stakes and Senate Properties office buil-ings, "Going with the grain" by Cathy Lang Ho
avs 3/2003, p46-47, Guinean projects, "Arkitekúr sem verður til úr p sem er til staðar" by Halldóra Arnardóttir
Nature 424, p718-720, 'Do you want to work here ?' by Laura Bonett p858-859, 'Bold architecture aims to...' by Kendall Powell, both articl referring to the Max Planck Institute of Molecular Cell Biology and G netics
Form Function Finland, p6-9, Stakes and Senate Properties office buil-ings
Yapi 261, p66-70, Media Center Lume
L'Arca 182, p54-63, Max Planck Institute of Molecular Cell Biology an Genetics, 'Tra natura e tecnologia' by Isabella Goldmann
L'industria delle costruzioni 370, p32-41, Media Center Lume, 'Centr di formazione multimediale, Università di Arte e Design a Helsinki' b Renato Morganti
Lotus 116, p60-71, Guinean projects, 'Architects and Antropologists' b Carolina Fois
Yapi 258, p80-87, Max Planck Institute of Molecular Cell Biology an Genetics
GA Houses 74 'Project 2003', p20-21, Villa Havina
AD Vol 73 1/2003 "Off the Radar", Guest edited by Brian Carter an Annette Le Cuyer, p 26-28, Kahere Eila Poultry Farming School
Architectural Record 1/2003, p110-117, Max Planck Institute of Mole-ular Cell Biology and Genetics, 'A Sense of Community for Scientist by David Cohn
Architektura 1/2003, p100, Max Planck Institute of Molecular Cell Biol-gy and Genetics, 'Blacha-schocly' by Barbara Lesnik

2002

Arkkitehti 6/2002, p28-37, Stakes and Senate Properties office building
Architecture 11/2002, p76-79, House Kosketus, 'Light Touch' by C.C Sullivan
Innovations in Steel 2002 / Educational Facilities, p21, Media Cente Lume
Wettbewerbe Aktuell 1/2002, p57, Plenarsaal des Bayerischen Landtag im Maximilianeum in München, competition entry
Architectural Review 6/2002, p40-45, Vuotalo Cultural Center, 'Cool Ur banity' by Henry Miles
Architectural Review 8/2002, p52-55, Max Planck Institute of Molecula Cell Biology and Genetics, 'Functional Biology' by Layla Dawson
md 8/2002, 27-31, Media Center Lume, 'A new Building for Arabianran ta' by Susanne Tamborini
Alvar Aalto Gesellschaft Bulletin 15 / Sommer 2002, Max Planck Insti-tute of Molecular Cell Biology and Genetics
yapi 8/2002, p61-68, Vuotalo Cultural Center
Architektur & Bauforum, 3-4/2002, p61, Juminkeko
Arkkitehti 2/2002, p32-41, Max Planck Institute of Molecular Cell Biol-ogy and Genetics

《建筑师》2002 年第 1 期，第 48 ~ 55 页，沃塔洛文化中心；
《a + u》2002 年第 2 期，第 121 页，阿迦汗奖，卡荷尔埃拉家禽养殖学校；
《居住》2002 年第 4 期，第 24 页，埃拉别墅，“芬兰人在几内亚”，莉蒂亚・李。

2001 年

《建筑师》2001 年第 12 期，第 74 ~ 87 页，“美是简单——芬兰建筑的现代传统”，沃尔夫冈・珍・斯托克；
《印度建筑师 & 建筑师》第 94 ~ 97 页，阿迦汗奖，卡荷尔埃拉家禽养殖学校；
《室内设计 + 建筑》2001 年第 9 期，第 42 ~ 47 页，“芬兰驻华盛顿大使馆的建筑风格”，方海；
《室内设计 + 建筑》2001 年第 10 期，第 50 ~ 54 页，“麦当劳芬兰总部大楼”，方海；
《区域》第 58 期，第 22 ~ 29 页，“埃拉别墅”，帕都・吉阿迪罗；
《建筑评论》2001 年第 11 期，第 58 ~ 59 页，卡荷尔埃拉家禽养殖学校，阿迦汗奖；
《建筑评论》2001 年第 9 期，第 62 ~ 64 页，卢米媒体中心，“教学媒体”，彼得・戴维；
《建筑实录》2001 年第 7 期，第 98 ~ 103 页，沃塔洛文化中心，彼得・麦克肯斯沃；
《建筑师》2001 年第 1 期，第 44 ~ 47 页，科斯克特斯预制房屋；
《方舟》157，第 48 ~ 55 页，卢米媒体中心，“元语言与建筑”，卡里罗・帕卡尼利；
《维度 V15——“限制”》2001 年，阿尔弗雷德陶布曼建筑学院及密歇根大学城市规划学院，第 118 ~ 121 页，“开始绘图——与米高・海基宁的讨论”。

2000 年

《建筑》2000 年第 12 期，第 104 ~ 115 页，卢米媒体中心，凯茜・朗・霍；
《芬兰形式功能》2000 年第 3 ~ 4 期，第 88 ~ 90 页，科斯克特斯预制房屋，“芬兰住宅”，米纳玛丽亚・科斯凯拉；
《建筑》2000 年第 106 期，第 39 ~ 40 页，科斯克特斯预制房屋；
《摘要》2000 年第 4 期，第 34 ~ 40 页，卢米媒体中心；
《对话》2000 年第 8 期，第 68 ~ 71 页，埃拉别墅；
《建筑师》2000 年第 3 期，第 40 ~ 47 页，卢米媒体中心，萨穆利・米提内评论，第 48 ~ 53 页，家禽农场学校；
《芬兰形式功能》2000 年第 1 期，第 20 ~ 25 页，卢米媒体中心，“电影是如今最重要的艺术！”，劳瑞・陶和宁。
《建筑艺术》2000 年第 27 ~ 28 期，第 13 ~ 27 页，卢米媒体中心和约翰霍普金斯大学学生艺术中心，“理性与经济的回归”，埃萨・拉克索宁；
《室内设计》2000 年 4 月，第 48 ~ 51 页，芬兰驻华盛顿大使馆，“后恐怖外交”，简・C.・勒夫勒；
《建筑》2000 年第 1 期，第 84 ~ 89 页，“Kolonihavehus”；
《Puu-lehti》2000 年第 1 期，第 6 ~ 11 页，尤米凯科；
《细节》2000 年第 1 期，Bauen mit Holz，第 72 ~ 76 页，尤米凯科；
《统计数据》2000 年第 5 期，第 13 页，尤米凯科，计算机时代的中世纪建筑文艺复兴。

Arkkitehti 1/2002, p48-55, Vuotalo Cultural Center
a+u 2/2002, p121, Aga Khan Awards, Kahere Eila Poultry Farming School
dwell 4/2002, p24, Villa Eila, ‘Finns in Guinea’ by Lydia Lee

2001

Architectur Aktuell 12/2001, p74-87, ‘Beauty is Simplicity - the Modern Tradition of Finnish Architecture’ by Wolfgang Jean Stock
Indian Architect & Builder, p94-97, Aga Khan Awards, Kahere Eila Poultry Farming School
Interiors Design+Construction 9/2001, p42-47, ‘The Architecture Style of the Finland Embassy in Washington’ by Fang Hai
Interiors Design+Construction 10/2001, p50-54, ‘The Building of McDonald’s General Headquarters in Finland’ by Fang Hai
Area 58, p22-29, Villa Eila, article by Pado Giardiello
Architectural Review 11/2001, p58-59, Aga Khan Awards, The Kahere Eila Poultry Farming School
Architectural Review 9/2001, p62-64, Media Center Lume, ‘Teaching Medium’ by Peter Davey
Architectural Record 7/2001, p98-103, Vuotalo Cultural Center by Peter McKeith
Arkkitehti 1/2001, p44-47, House Kosketus
L’Arca 157, p48-55, Media Center Lume, ‘Metalinguaggio e architettura’ by Carlo Paganelli
Dimensions V15, ‘Limits’, 2001, Journal of the Alfred Taubman College of Architecture + Urban Planning University of Michigan, p118-121, ‘Start Drawing’, a discussion with Mikko Heikkinen

2000

Architecture 12/2000, p104-115, Media Center Lume by Cathy Lang Ho
Form Function Finland 3-4/2000, p88-90, House Kosketus, ‘Finnish Homes’ by Minnamaria Koskela
d’Architectures 106/2000, p39-40, House Kosketus
Abstract 4/2000, p34-40, Media Center Lume
Dialogue 8/2000, p68-71, Villa Eila
Arkkitehti 3/2000, p40-47, Media Center Lume, criticism by Samuli Miettinen, p48-53, School for Poultry Farmers
Form Function Finland 1/2000, p20-25, Media Center Lume, ‘Film is now the most important of the arts !’ by Lauri Törhönen
Ehituskunst 27-28/2000, p13-27, Media Center Lume and Johns Hopkins University Student Art Center, ‘The return of Reason and Economy’ by Esa Laaksonen
Interiors April 2000, p48-51, Finnish Embassy in Washington, ‘Post Terror Diplomacy’ by Jane C. Loeffler
Architecture 1/2000, p84-89, Kolonihavehus
Puu-lehti 1/2000, p6-11, Juminkeko
Detail 1/2000 Bauen mit Holz, p72-76, Juminkeko
Staty bu 5/2000, p13, Juminkeko, Medinés Architekturos renesansas kompiuteriu eroje

1999 年

《建筑用钢》1999 年第 151 期，第 26 ~ 27 页，瓦萨里球场入口设计；
《阿格斯》第 10 期，第 25 ~ 26 页，瓦萨里球场入口设计；
《建筑实录》1999 年第 3 期，第 90 ~ 93 页，麦当劳总部第，彼得·马肯斯；
《建筑师》1999 年第 1 期，第 56 ~ 59 页，天使工作室；
《建筑》第 315 期，第 46 ~ 49 页，应急服务学院；
《德国建筑杂志》1999 年第 7 期，第 39 ~ 44 页，麦当劳总部；
《辑要 +》第 34 期，第 80 ~ 85 页，埃尔南·巴贝罗·萨则博和塞尔吉奥·D·卡斯蒂格里奥尼，芬兰罗瓦涅米机场，“在芬兰着陆”；
《建筑师》1999 年第 6 期，第 34 ~ 39 页，尤米凯科，马库斯·阿托尼的评论，第 64 ~ 67 页，瓦萨里球场入口设计；
《建筑》1999 年第 526 期（总 526 期），第 429 ~ 452 页，天使工作室、军官培训单位库奥皮奥、麦当劳总部、艺术与设计大学视听中心、罗瓦涅米机场航站楼、马克斯普朗克分子生物学和遗传学研究院。

1998 年

《纪念碑》1998 年第 24 期，第 50 ~ 61 页，“投入少，收获多”，彼得·马肯斯；
《细节》1998 年第 1 期，第 56 ~ 60 页，几内亚卫生中心；
《建筑师》1998 年第 1 期，第 34 ~ 41 页，麦当劳总部，赖纳·曼拉玛奇评论；
《建筑世界》1998 年第 10 期，第 460 ~ 465 页，“芬兰建筑”，克里斯托夫·阿芬传格；
《标准杂志》1998 年第 10 期，第 22 ~ 25 页，“芬兰建筑师阿尔瓦·阿尔托的脚步——走在潮湿的苔藓上”，马克斯·伯卡；
《斯堪的纳维亚评论》第 66 ~ 67 页，“边境庄园：填满边界”，威廉·摩根撰写；
《建设者》1998 年第 5 期，第 6 页，麦当劳总部；
《建筑师》1998 年第 9 期，第 24 页，麦当劳总部，“麦当劳赫尔辛基总部”，林恩·达曼·伦德；
《美丽的家》第 657 期，第 28 ~ 33 页，“库奥皮奥专业培训中心”；
《建筑评论》1998 年 8 月，第 70 ~ 73 页，麦当劳总部，“品牌的典范”，亨利·迈尔斯；
《建筑师》1998 年第 4 期，第 26 ~ 33 页，村卫生所、埃拉别墅、库塔市和波都考拉的小学，以及养鸡户培训学校；
《论坛》1998 年第 3 期，第 34 ~ 37 页，“麦当劳”，乌尔夫·瑞斯托姆；
《建筑》1998 年第 10 期，第 158 ~ 160 页，“授奖区”，埃里克·亚当斯；
《芬兰形式功能》1998 年第 3 期，第 8 ~ 10 页，麦当劳总部；
《多莫斯》第 810 期，第 26 ~ 31 页，麦当劳总部；
《建筑师》1998 年第 6 期，第 62 ~ 67 页，应急服务学院；
《建筑业》第 321 期，第 66 页，特伯利服务站；
《生活建筑》第 72 期，第 90 ~ 93 页，埃拉别墅；
《建筑 + 比赛》第 176 期，第 32 ~ 33 页，麦当劳总部；
《建筑艺术》1998 年第 22 ~ 23 期，第 38 ~ 45 页，麦当劳总部。

1999

Bouwen met Staal 151/1999, p26-27, Vuosaari Gateway
Argus 10, p25-26, Vuosaari Gateway
Architectural Record 3/1999, p90-93, McDonald's headquarters, articl by Peter McKeith
Arkkitehti 1/99, p56-59, Angel Studios
Arquitectura 315, p46-49, Emergency Services College
Deutsche Bauzeitschrift 7/99, p39-44, McDonald's Headquarters
Summa+ 34, p80-85, Rovaniemi Airport 'Aterrzaje en Finlandia' b Hernán Barbero Sarzabal and Sergio D. Castiglioni
Arkkitehti 6/99, p34-39, Juminkeko, criticism by Markus Aaltonen, p64 67, Vuosaari Gateway
L'architettura 526/526 1999, p429-452, Angel Studios, Officers training unit Kuopio, McDonald's headquarters, Audiovisual Center of the University of Art and Design, Airport terminal, Rovaniemi, Max Planck Institut for Molecular Biology and Genetics

1998

Monument 24 1998, p50-61, 'Making more with less' by Peter MacKeith
Detail 1/1998, p56-60, Health Center in Guinea
Arkkitehti 1/1998, p34-41, McDonald's Headquarters, criticism by Rainer Mahlamäki
Bauwelt 10/1998, p460-465, 'Finnische Architektur' by Christoph Affentranger
De Standaard Magazine 10/1998, p22-25, 'Finse architectuur in de voetsporen van Alvar Aalto - Wandelen over nat mos' by Max Borka
Scandinavian Review p66-67, 'Frontier Manor: Filling up at the Border' by William Morgan
Baumeister 5/1998, p6, McDonald's Headquarters
Arkitekten 9/1998, p24, McDonald's Headquarters, 'McDonalds hovedsæde i Helsinki' by Lene Dammand Lund
Casabella 657, p28-33, 'Centro di formazione professionale a Kuopio'
The architectural Review August 1998, p70-73, McDonald's Headquarters, 'Apotheosis of a Brand' by Henry Miles
Arkkitehti 4/1998, p26-33, Village Health Unit, Villa Eila, Elementary schools in Madina Kouta and Boundou Koura, School for chicken farmers
Forum 3/1998, p34-37, 'McDonald's' by Ulf Ringström
Architecture 10/1998, p158-160, 'Winner's Circle' by Eric Adams
Form Function Finland 3/1998, p8-10, McDonald's Headquarters
Domus 810, p26-31, McDonald's Headquarters
Arkkitehti 6/1998, p62-67, Emergency Services College
L'industria delle Construzioni 321, p66, Teboil Service Station
Arquitectura Viva 72, p90-93, Villa Eila
Architektur + Wettbewerbe 176, p32-33, McDonald's Headquarters
Ehituskunst 22-23/1998, p38-45, McDonald's Headquarters

997 年

斯堪的纳维亚的评论》1997 年，第 59 ~ 64 页，“库伦尼哈文：
示折中主义的地方”，玛丽亚・史蒂格利茨；
今天的建筑》第 309 期，第 22 ~ 27 页，保健中心和埃拉别墅，
几内亚马里的药房和别墅”，阿梅勒・拉瓦卢；
建筑设计“建筑之光”》，第 82 ~ 87 页，芬兰驻华盛顿大使馆；
纪念碑》1997 年第 18 期，第 86 ~ 89 页，芬兰驻华盛顿大使馆；
空间设计》1997 年第 10 期，第 43 ~ 45 页，芬兰驻华盛顿大使馆；
a + u》第 326 期，第 40 ~ 47 页，埃拉别墅；
八角形》第 125 期，第 46 ~ 48 页，赫尔辛基艺术设计大学视
中心及天使工作室。

996 年

芬兰形式功能》1996 年第 1 期，第 36 ~ 41 页，赫尔辛基艺
设计大学视听中心，“一个大梦想”，简・沃维基；
建筑师》1996 年第 2 期，第 42 页，“丰富的简约”，克里斯托弗・哈
，《专著评论》，古斯塔沃・吉利；
建筑师》1996 年第 2 ~ 3 期，第 32 ~ 35 页，几内亚卫生中心，
'Fulanitalontyömaalla'”，西莫・弗雷泽；
建筑师》1996 年第 4 期，第 42 ~ 45 页，麦当劳总部，第
~ 49 页，特伯利服务站；
建筑师》1996 年第 5 ~ 6 页，第 52 ~ 55 页，瓦萨里球场入口设计；
大都市》1996 年第 4 月，第 62 ~ 65 页，迈克尔・韦伯，“北
星——芬兰设计”，参考芬兰驻华盛顿大使馆；
OZ 卷》1996 年第 18 期，第 18 ~ 23 页，特伯利服务站，“超越
学思想的建筑”，米高・海基宁；
Glasforum》1996 年第 2 期，第 3 ~ 6 页，芬兰驻华盛顿大使馆，
芬兰驻华盛顿大使馆”，萨宾尼・威辛格；
铰链》第 27 期，“世界之巅的斯堪的纳维亚建筑”，第 20 ~ 21 页；
多莫斯》第 786 期，第 10 ~ 17 页，几内亚项目；
建筑》1996 年第 10 期，第 44 页，库伦尼哈文建筑公园，“建
师更新丹麦避暑别墅”，内德・克莱默；
设计书评》第 37 ~ 38 期，第 53 页，老年人住房和娱乐中心；
阿奎斯》1996 年第 11 期，第 42 ~ 47 页，应急服务学院。

995 年

建筑师》1995 年第 1 期，第 26 ~ 31 页，“来自芬兰的建筑”；
建设者》1995 年第 1 期，第 30 ~ 35 页，应急服务学院，“库
皮奥救援服务学校”，泰伯・约肯宁；
房子》1995 年第 2 期，第 10 页，芬兰驻华盛顿大使馆，“芬
的‘玻璃’外交”，朱莉安娜・巴林特；
建筑与商业》1995 年第 3 期，第 12 ~ 13 页，芬兰驻华盛顿
使馆，“芬兰在美国”，马尔桑・达尔奇克；
芬兰设计》1995 年，第 6 ~ 12 页，“大使馆成为华盛顿的亮
建筑”，韦夫・斯腾格；

1997

Scandinavian Review 1997, p59-64, ‘Kolonihaven: A Showcase for Eclecticism’ by Maria Stieglitz
L’architecture d’aujourd’hui 309, p22-27, Health Center and Villa Eila, ‘Dispensaire et villa à Mali, Guinée’ by Armelle Lavalou
Architectural Design ‘Light in Architecture’, p82-87, Finnish Embassy in Washington
Monument 18/1997, p86-89, Finnish Embassy in Washington
Space Design 9710, p43-45, Finnish Embassy in Washington
a+u 326, p40-47, Villa Eila
Ottagono 125, p46-48, Audiovisual Center of UIAH, Angel Studios

1996

Form Function Finland 1/1996, p36-41, Audiovisual Center for UIAH, ‘A Big Dream’ by Jan Verwijen
Arkitekten 2/1996, p42, ‘Beriget enkelhed’ by Christoffer Harlang, review of the monograph by Gustavo Gili
Arkkitehti 2-3/1996, p32-35, Health Center in Guinea, ‘Fulanitalon työmaalla’ by Simo Freese
Arkkitehti 4/1996, p42-45, McDonald’s Headquarters, p46-49, Teboil Service Station
Arkkitehti 5-6/1996, p52-55, Vuosaari Gateway
Metropolis April 1996, p62-65, ‘Arctic Stars - Finnish Design’ by Michael Webb, ref. Finnish Embassy in Washington
OZ Volume 18/1996, p18-23, Teboil Service Station, ‘Architecture Beyond Philosophical Ideas’ by Mikko Heikkinen
Glasforum 2/1996, p3-6, Finnish Embassy in Washington, ‘Finnische Botschaft Washington’ by Sabine Weissinger
Hinge 27, ‘Scandinavian Architecture on the Top of the World’, p20-21
Domus 786, p10-17, Guinean projects
Architecture 10/1996, p44, Kolonihaven Architecture Park, ‘Architects Update Danish Summer Houses’ by Ned Cramer
Design Book Review 37-38, p53, Senior Citizen Housing and Amenity Center
Arquis 11/1996, p42-47, Emergency Services College

1995

Der Architekt 1/1995, p26-31, ‘Architektur aus Finnland’
Baumeister 1/1995, p30-35, Emergency Services College, ‘Schule für Rettungsdienste in Kuopio’ by Teppo Jokinen
Häuser 2/1995, p10, Finnish Embassy in Washington, ‘Finnland Für ‘Gläserne’ Diplomatie’ by Juliana Balint
Arkitektura & Biznes 3/1995, p12-13, Finnish Embassy in Washington, ‘Finlandia w Ameryce’ by Marcin Wlodarczyk
Design in Finland 1995 p6-12, ‘Embassy takes the Washington spotlight’ by Wif Stenger

《建筑业》第 281 期，第 20 ～ 39 页，芬兰驻华盛顿大使馆，公民住房和娱乐中心，欧洲电影学院，“海基宁与科莫宁：90 年代的三件作品”，雷纳托・莫尔甘蒂；
《新苏黎世报（NZZ）对开本》1995 年第 6 期，第 78 ～ 79 页，芬兰驻华盛顿大使馆，绿色信息，罗曼・何雷斯坦；
《项目》第 185 期，第 46 ～ 48 页，“罗瓦涅米机场”，伊娃・詹森；
《建筑师》1995 年第 2 ～ 3 期，第 56 ～ 61 页，老年人住房和娱乐中心；
《室内设计》第 44 期，第 38 ～ 47 页，“芬兰驻华盛顿大使馆——记忆中的风景”，索雷达・洛伦佐；
《建筑新闻》第 184 期，第 52 ～ 63 页，芬兰驻华盛顿大使馆，“在华盛顿中心的芬兰灵魂”，威廉・摩根；
《细节》1995 年 5 月，第 845 ～ 850 页，芬兰驻华盛顿大使馆；
《建筑生活》第 55 期，“斯堪的纳维亚人”，第 94 ～ 97 页，欧洲电影学院；
《技术与建筑》第 422 期，第 96 ～ 101 页，芬兰驻华盛顿大使馆，“北欧精神”，比拉特丽斯・赫辛拉；
《建筑设计“建筑的力量”》，第 50 ～ 51 页，芬兰驻华盛顿大使馆。

1994 年

《多莫斯》第 759 期，20 ～ 25 页，欧洲电影学院；
《技术与建筑》第 414 期，第 26 ～ 31 页，欧洲电影学院，“宽敞明亮的房间”，玛丽—克里斯汀・洛里耶；
《建筑评论》1994 年 6 月，第 78 ～ 83 页，欧洲电影学院，“冻结框架”，亨利・迈尔斯；
《建筑师》1994 年第 2—3 期，第 22 ～ 31 页，芬兰驻华盛顿大使馆；
《北欧建筑研究》1994 年第 3 期，第 65 ～ 74 页，克莱斯・卡尔登比的“理论在实践中：对批评边界的反思”，参考芬兰科学中心；
《AIT》1994 年第 7 ～ 8 期，第 58 ～ 63 页，欧洲电影学院，“电影 ab”，迪特马尔・丹纳；
《芬兰形式功能》1994 年第 3 期，第 4 ～ 9 页，芬兰驻华盛顿大使馆，“一座建筑的融合”，卡里娜・尤赫拉—希特沙罗；
《先进的建筑》，1994 年 7 月，第 74 ～ 83 页，“宇宙的连接——海基宁和科莫宁的建筑”，威廉・摩根；
《建筑评论》，1994 年 10 月，第 36 ～ 42 页，芬兰驻华盛顿大使馆，“外交社区”，威廉・摩根；
《建筑评论》，1994 年 11 月，第 60 ～ 67 页，芬兰驻华盛顿大使馆，“外交手段”，克利福德・A・皮尔森；
《MD》1994 年第 9 期，第 63 ～ 64 页，拉赫蒂住房交易会的儿童小屋，“父母无法访问”，朱莉安娜・巴林特；
《建筑世界》1994 年第 40 ～ 41 期，第 2288 ～ 2291 页，芬兰驻华盛顿大使馆；
《美丽的家》第 617 期，第 54 ～ 59 页，“芬兰驻华盛顿大使馆”，赛巴斯蒂亚诺・巴瑞多里尼；
《新桥》第 169 期，第 10 ～ 16 页，“米高・海基宁和马尔库・科莫宁：芬兰的新建筑”，雷蒙德・瑞恩；
《斯堪的纳维亚的评论》，第 76 ～ 81 页，“在华盛顿中心的芬兰灵魂”，威廉・摩根。

L'industria delle costruzioni 281, p20-39, Finnish Embassy in Washir ton, Senior Citizen Housing and Amenity Center, European Film C lege, 'Heikkinen e Komonen, tre opere degli anni novanta' by Rena Morganti
NZZ Folio 6/1995, p78-79, Finnish Embassy in Washington, 'Die grü Botschaft' by Roman Hollenstein
Projeto 185, p46-48, 'Aeroporto de Rovaniemi' by Eva Janson
Arkkitehti 2-3/1995, p56-61, Senior Citizen Housing and Amenity Cent
Diseño Interior 44, p38-47, 'Embajada de Finlandia en Washington - paisaje en la memoria' by Soledad Lorenzo
Architektur Aktuell 184, p52-63, Finnish Embassy in Washington, 'D Seele Finnlands in Herzen Washingtons' by William Morgan
Detail 5/1995, p845-850, Finnish Embassy in Washington
Arquitectura Viva 55, 'Scandinavians', p94-97, European Film College
Techniques et Architecture 422, p96-101, Finnish Embassy in Washin ton, 'Esprit nordique' by Beatrice Houzelle
Architectural Design 'The Power of Architecture', p50-51, Finnish Embas in Washington

1994

Domus 759, p20-25, European Film College
Techniques et Architecture 414, p26-31, European Film College, 'Chai bres claires' by Marie-Christine Loriers
Architectural Review June 1994, p78-83, European Film College 'Free Frame' by Henry Miles
Arkkitehti 2-3/1994, p22-31, Finnish Embassy in Washington
Nordisk arkitektur forskning 3/1994, p65-74, 'Teorin i praktiken, ref ktionen kring kritikens gränser' by Claes Caldenby ref. Finnish Scien Center
AIT 7-8/1994, p58-63, European Film College, 'Film ab' by Dietmar Da ner
Form Function Finland 3/1994, p4-9, Finnish Embassy in Washington ' Building that Blends in' by Katriina Jauhola-Seitsalo
Progressive Architecture, July 1994, p74-83, 'The Cosmic Connectio the Architecture of Heikkinen and Komonen' by William Morgan
Architectural Review October 1994, p36-42, Finnish Embassy in Was ington, 'Diplomatic Community' by William Morgan
Architectural Record November 1994, p60-67, Finnish Embassy in Was ington, 'Diplomatic Maneuvers' by Clifford A. Pearson
MD 9/1994, p63-64, Children's Cabins at Lahti Housing Fair, 'F Elterns kein Zutritt' by Juliana Balint
Bauwelt 40-41/1994, p2288-2291, Finnish Embassy in Washington
Casabella 617, p54-59, 'Ambasciata di Finlandia a Washington D.C.' by S bastiano Brandolini
Neuf-Nieuw 169, p10-16, 'Mikko Heikkinen & Markku Komonen, l nouvelle architecture finlandaise' by Raymund Ryan
Scandinavian Review, p76-81, 'The Soul of Finland in the Heart of Was ington' by William Morgan

993年

《建筑师》1993年第1期，第24～33页，应急服务学院，尤哈·耶克莱宁评论；

《建筑实录》1993年2月，第40页，"走向光明"，克利福德·皮森谈关于最近的芬兰建筑；

《建筑评论》1993年3月，第52～54页，芬兰驻华盛顿大使馆，"精美芬兰式建筑"，彼得·戴维；

《建筑师》1993年第5期，第198页，欧洲电影学院，"丹麦项目"，夫·莱尔·索伦森；

《建筑师》1993年第3期，第44～45页，几内亚卫生中心；

《建筑评论》1993年4月，第81页，在吕伐登举办的北欧艺术建筑展览的芬兰馆，"视觉声学"，彼得·戴维；

《架构和布局》，第74～78页，"芬兰设计的破产"，托尔达·比亚纳松；

《建筑生活》第30期，芬兰建筑特刊，彼得·戴维、斯科特·普和威廉·柯蒂斯撰写，第38～41页，应急服务学院，第2～45页，罗瓦涅米机场；

《MagyarÉpitömüvészet》1993年第6期，第18～21页，罗瓦米机场；

《建筑师》1993年第6期，第72～81页，欧洲电影学院，"大和野蛮的盒子"，拉尔斯·蒂斯·克努森评论；

《建筑评论》1993年8月，第57～61页，应急服务学院，"紧芬兰"，彼得·马肯斯；

《美丽的家》1993年第8期，第37～41页，努克文化中心；

《建筑评论》1993年9月，第67～71页，罗瓦涅米机场，"光"，伊娃·詹森；

《建筑业》第264期，第26～37页，应急服务学院和罗瓦涅米场，"两个租最近的作品"，雷纳托·莫尔甘蒂；

《丹麦建筑师》1993年第7期，第284～293页，欧洲电影学院，"极简主义的盒子"，简·W·汉森评论；

《方舟》第73期，托德·达兰关于FTL / Happold作品的"结构节"，参考芬兰大使馆的天棚；

《建筑＋比赛》第156期，第22～23页，应急服务学院。

992年

《芬兰形式功能》1992年第1期，第60～65页，"建筑与艺术：兰合作"，彼得·马肯斯；

《报告》1992年第2～3期，第22～23页，罗瓦涅米机场，"两通往自然之路"，卡瓦·泰帕莱迪；

《建筑师》1992年第4～5期，第52～59页，罗瓦涅米机场，迈娅·凯拉莫评论；

《建筑实录》1992年5月，第24页，芬兰驻华盛顿大使馆；

《进步建筑杂志》1992年5月，第154～155页，"外国简介——兰"，斯科特·普尔谈关于近代芬兰建筑；

《美丽的家》第593期，第4～17页，"海基宁与科莫宁最近的作品—科学与艺术交汇处的建筑"，尤哈·埃罗恩；

《竞赛》1992年第2期，第42～49页，"芬兰的竞赛世纪"，威廉·摩根；

《建筑设计1095》，第10页，芬兰驻华盛顿大使馆，"粘贴王冠中的宝石"，约翰·威尔士；

1993

Arkkitehti 1/1993, p24-33, Emergency Services College, criticism by Juha Jääskeläinen

Architectural Record February 1993, p40, 'Following the Sun' by Clifford Pearson on recent Finnish architecture

Architectural Review March 1993, p52-54, Finnish Embassy in Washington, 'Fine Finnish' by Peter Davey

Arkitekten 5/1993, p198, European Film College, 'Danske projekter' by Leif Leer Sørensen

Arkkitehti 3/1993, p44-45, Health Center in Guinea

Architectural Review April 1993, p81, Finnish Pavilion for a Nordic Art and Architecture Exhibition in Leeuwarden, 'Visual Acoustics' by Peter Davey

Arkitektur og skipulag, p74-78, 'Finnsk hönnun á framabraut' by Gudjón Bjarnason

Arquitectura Viva 30, Special issue on Finnish architecture with articles by Peter Davey, Scott Poole and William Curtis, p38-41, Emergency Services College, p42-45, Rovaniemi Airport

Magyar Épitömüvészet 6/1993, p18-21, Rovaniemi Airport

Arkkitehti 6/1993, p72-81, European Film College, criticism by Lars Thiis Knudsen 'Bold and Brutal Boxes'

Architectural Review August 1993, p57-61, Emergency Services College, 'Emergency Finnish' by Peter MacKeith

Ehituskunst 8/1993, p37-41, Cultural Center in Nuuk

Architectural Review September 1993, p67-71, Rovaniemi Airport, 'Circle of Light' by Eva Janson

L'industria delle costruzioni 264, p26-37, Emergency Services College and Rovaniemi Airport, 'Due opere recenti' by Renato Morganti

Arkitektur DK 7/1993, p284-293 , European Film College, 'Minimalistiske kasser' criticism by Jan W. Hänsen

L'Arca 73, 'Structural Detailing' by Todd Dalland on the works of FTL / Happold ref. the canopy of the Finnish Embassy

Architektur + Wettbewerbe 156, p22-23, Emergency Services College

1992

Form Function Finland 1/1992, p60-65, 'Architecture + Art, Finnish Collaborations' by Peter MacKeith

Report 2-3/1992, p22-23, Rovaniemi Airport 'Two Roads to Nature' by Kaarin Taipale

Arkkitehti 4-5/1992, p52-59, Rovaniemi Airport, criticism by Maija Kairamo

Architectural Record May 1992, p 24, Finnish Embassy in Washington

Progressive Architecture May 1992, p154-155, 'Foreign brief: Finland' by Scott Poole on recent Finnish architecture

Casabella 593, p4-17, 'Opere recenti di Heikkinen - Komonen' essay by Juha Ilonen: 'Architecture at the intersection of science and art'

Competitions 2/1992, p42-49, 'A Century of Competitions in Finland' by William Morgan

Building Design 1095, p10, Finnish Embassy in Washington, 'Jewel in a paste crown' by John Welsh

《建筑生活》第27期，第65页，“马德里的愿景——五项建议”，玛莎·索恩；
《芬兰的形式与功能——西班牙语的特殊号码》，1992年，第82～85页，“罗瓦涅米机场”，安妮卡·尼伯格；
《Baf nytt》1992年第2期，第6页，罗瓦涅米机场；
《建筑师》1992年第7期，第81～83页，Matrix H_2O，马德里。

Arquitectura Viva 27, p 65, Visiones para Madrid, Cinco propuest foráneas' by Martha Thorne
Form Function Finland, Número especial en español 1992, p82-85, 'I Aeropuerto de Rovaniemi' by Annika Nyberg
Baf nytt 2/1992, p6, Rovaniemi Airport
Arkkitehti 7/8 1992, p81-83, Matrix H2O Madrid

1991年

《阿基斯》1991年第1期，第21～33页，“芬兰文化宫”，汉斯·德·摩尔；
《建筑师》1991年第1期，第82～84页，北欧艺术和建筑展览的芬兰馆，吕伐登；
《建筑师》1991年第4～5期，第35～38页，芬兰驻华盛顿大使馆，第39～43页，欧洲电影学院；
《建筑世界》1991年第5期，第166～167页，芬兰科学中心；
《建筑业》第233期，第26～33页，“芬兰科学中心”，雷纳托·莫尔甘蒂；
《设计日志》1991年第43期，第16页，芬兰科学中心，尼塔·尼库拉；
《建筑师》1991年第17期，第510～523页，欧洲电影学院竞赛评论；
《美丽的家》第585期，第36～38页，捷克首都布拉格研讨会，1991年。

1991

Archis 1/1991, p21-33, 'Cultuurpaleizen in Finland' by Hans de Moor
Arkkitehti 1/1991, p82-84, Finnish Pavilion for a Nordic Art and Arch tecture Exhibition in Leeuwarden
Arkkitehti 4-5/1991, p35-38, Finnish Embassy in Washington; p39-4 European Film College
Bauwelt 5/1991, p166-167, Finnish Science Center
L'industria delle costruzioni 233, p26-33, 'Il Centro Finlandese delle Sc enza' by Renato Morganti
Design Journal 43/1991, p16, Finnish Science Center, article by Riitt Nikula
Arkitekten 17/1991, p510-523, Competition Review on European Fil College
Casabella 585, p36-38, Workshop Prague 1991

1990年

《方舟》第36期，第8～17页，“尤里卡”，朱莉安娜·巴林特；
《建筑师》1990年3月，第31～37页，“万塔的芬兰科学博物馆”，卡雷尔·范·布吕亨；
《建筑评论》1990年3月，第30～36页，“赫尤里卡元素”，彼得·戴维的评论；
《植物、建筑与生活》1990年第4期，第12~15页，芬兰科学中心；
《细节》1990年第4期，第389～394页，芬兰科学中心；
《建筑师》1990年第6期，第56～59页，赫尤里卡桥；第40、41页，罗瓦涅米机场；
《创造建筑》1990年8月至9月，第138～141页，“赫尤里卡”，M.H. 康塔尔；
《生活建筑》第9期，第94～103页，芬兰科学中心；
《a + u》1990年第7期，第7～29页，芬兰科学中心，罗杰·康纳评论；
《建筑设计》1990年12月，罗瓦涅米机场，“克劳斯和效应”，约翰·威尔士；
《芬兰形式》1999年，第26～33页，芬兰科学中心，“问正确问题的艺术”，里斯托·皮特卡南。

1990

L'Arca 36, p8-17, 'Eureka' by Juliana Balint
de Architect March 1990, p31-37, 'Fins Wetenschapsmuseum te Vantaa by Carel van Bruggen
The Architectural Review March 1990, p30-36, 'Heureka Elements', crit icism by Peter Davey
Werk, Bauen + Wohnen 4/1990, p12-15, Finnish Science Center
Detail 4/1990, p389-394, Finnish Science Center
Arkkitehti 6/1990, p56-59, Heureka-bridge; p40-41, Rovaniemi Airport
Architecture interieure crée August-September 1990, p138-141, 'Heureka by M.H. Contal
Living Architecture 9, p94-103, Finnish Science Center
a+u 7/1990, p7-29, Finnish Science Center, criticism by Roger Connah
Building Design December 1990, Rovaniemi Airport, 'Claus and effect by John Welsh
Formes Finlandaises 1990 p26-33, Finnish Science Center, 'L'art de pos er la bonne question' by Risto Pitkänen

1989年

《建筑设计》第944期，第16～17页，“芬兰风格”，约翰·威尔士；
《建筑》1989年9月，第58～61页，“阿尔托时代的最新作品”，简妮·贝内特；

1989

Building Design 944, p16-17, 'Finnish in Style' by John Welsh
Architecture September 1989, p58-61, 'Recent Works of the Post-Aalt Generation' by Janey Bennet

德国建筑报纸》1989 年 9 月，第 30 ~ 33 页，芬兰科学中心；
建筑师》1989 年第 4 期，第 44 ~ 57 页，芬兰科学中心；
芬兰形式功能》1989 年第 3 期，第 28 ~ 33 页，“赫尤里卡——个学习的经验积累”，埃娃・斯尔塔沃瑞；
建设者》1989 年第 10 期，第 44 ~ 45 页，芬兰科学中心；
美丽的家》第 562 期，第 29 页，芬兰科学中心。

Deutche Bauzeitung September 1989, p30-33, Finnish Science Center
Arkkitehti 4/1989, p44-57, Finnish Science Center
Form Function Finland 3/1989, p28-33, 'Heureka, a learning experience' by Eeva Siltavuori
Baumeister 10/1989, p44-45, Finnish Science Center
Casabella 562, p29, Finnish Science Center

987 年

1987

美丽的家》第 535 期，第 56 ~ 63 页，“芬兰科学中心”，塞斯蒂亚诺・巴瑞多里尼；
建筑师》1987 年第 7 期，第 32 ~ 37 页，洛玛・涅米拉假日酒店；
斯卡拉》第 11 期，第 6 页，芬兰科学中心；
芬兰的形式与功能》1987 年第 4 期，第 18 ~ 23 页，“赫尤卡——科学建筑”，卡林・泰帕尔。

Casabella 535, p56-63, 'Il centro finlandese della scienza' by Sebastiano Brandolini
Arkkitehti 7/1987, p32-37, Vacation Hotel LomaNiemelä
Skala 11, p 6, Finnish Science Center
Form Function Finland 4/87, p18-23, 'Heureka, Architecture of Science' by Kaarin Taipale

986 年

1986

建筑师》1986 年第 7 期，第 48 ~ 49 页，乌托莱塔周末别墅。

Arkkitehti 7/1986, p48-49, Weekend Cottage Uutturanta

985 年

1985

建筑师》1985 年第 2 期，第 63 ~ 65 页，帕尼基里纳改造，坦佩雷。

Arkkitehti 2/1985, p63-65, Renovation of Pyynikinlinna, Tampere

本书中介绍的模型，均由海基宁—科莫宁建筑事务所的设计师制作完成

The Models presented in this book are made at Heikkinen-Komonen Architects by

高・海基宁、珍妮・肯塔拉（系统生物学中心，设计工作）、尔科・阿尔蒂（瑟拉彻斯博物馆扩建项目及格陵兰国家美术馆景观模型）、安西・坎科宁（萨文林纳主图书馆的木制模型）、萨・鲁斯凯佩（萨文林纳主图书馆的木制模型、瑟拉彻斯博物扩建项目及格陵兰国家美术馆的景观模型）。

Mikko Heikkinen, Janne Kentala (Centre for Systems Biology, design), Erkko Aarti (Serlachius Museum Extension; National Gallery of Greenland, landscape model), Anssi Kankkunen (Savonlinna Main Library, wooden model), Esa Ruskeepää (Savonlinna Main Library, wooden model; Serlachius Museum Extension; National Gallery of Greenland, landscape model)

普罗・拉贾科斯基制作了系统生物学中心的模型，并提供了弗拉诺基奥住宅区和珀赫拉住宅区的模型组件，以及芬兰驻东京使馆的最终模型。梅里恩公路项目的亚克力材料部分由赫尔辛工业设计大学（UIAH）工作室制造完成。

Seppo Rajakoski has manufactured the model of The Centre for Systems Biology and provided components for the models of The Housing Block Flooranaukio, The Perhelä Housing Block and for the final model of The Embassy of Finland in Tokyo. The acrylic elements for The Merrion Road Project are manufactured at the UIAH (University of Industrial Design Helsinki) Workshop.

照片的版权所有（除非另有说明，否则均为海基宁—科莫宁建筑事务所所有）

Photo credits (H-K Architects, unless otherwise stated)

托布尔公园：米兰・罗勒（项目照片）
文林纳主图书馆：托马斯・乌西莫（项目照片）
洛拉诺基奥住宅区：尤西・第艾宁（项目照片）
斯托姆服务中心：托马斯・乌西莫（项目照片）
里达别墅：尤西・第艾宁（项目照片）
门林纳省档案馆：尤西・第艾宁（项目照片）
兰大使馆：尤西・第艾宁（项目照片）

Schönbühl Park: Milan Rohrer (project photo)
Savonlinna Main Library: Tuomas Uusheimo (project photo)
Housing Block Flooranaukio: Jussi Tiainen (project photo)
Lindström Service Centre: Tuomas Uusheimo (project photo)
Villa Frida: Tuomas Uusheimo (project photo)
Hämeenlinna Provincial Archive: Jussi Tiainen (project photos)
Embassy of Finland: Jussi Tiainen (project photo, model photos page 101)

作者简介 | About Author

米高·海基宁
1949 年生

1975 年　获赫尔辛基大学建筑硕士
1974 年　海基宁—科莫宁建筑事务所合伙人
2003—2008 年　芬兰艺术委员会提名建筑艺术教授
2010—2015 年　阿尔托大学“建筑与理论基础”教授
2005 年　获美国建筑师学会荣誉会员

马克库·科莫宁
1945 年生

1974 年　海基宁—科莫宁建筑事务所合伙人
1977—1980 年　Arkkitehti 杂志主编
1978—1986 年　芬兰建筑博物馆展览部主任
1992—2010 年　赫尔辛基科技大学建筑学教授 / 阿尔托大学教授
2005 年　获美国建筑师协会荣誉会员

珍妮·肯塔拉
1961 年生

1989 年　赫尔辛基科技大学建筑学硕士
1985 年—　海基宁—科莫宁建筑事务所建筑师
2010 年—　海基宁—科莫宁建筑事务所合伙人

马库·普马拉
1964 年生

2005 年　赫尔辛基工业大学建筑学硕士
1996 年—　海基宁—科莫宁建筑事务所建筑师
2010 年—　海基宁—科莫宁建筑事务所合伙人与 CEO

Mikko Heikkinen

b. 1949

Master of Science in Architecture, Helsinki University of Technology 1975

Heikkinen – Komonen Architects, partner 1974–

Artist Professor for Architecture nominated by the Arts Council of Finland 2003–2008

Professor of Basics of Architecture and Theory at Aalto University, Helsinki, 2010–2015

Honorary Fellow of the American Institution of Architects 2005–

Markku Komonen

b. 1945

Master of Science in Architecture, Helsinki University of Technology 1974

Heikkinen – Komonen Architects, partner 1974–

Editor-in-chief of Arkkitehti Magazine 1977–1980

Director of the Exhibition Department, Museum of Finnish Architecture 1978–1986

Professor of Architecture at Helsinki University of Technology / Aalto University 1992–2010

Honorary Fellow of the American Institution of Architects 2005–

Janne Kentala

b. 1961

Master of Science in Architecture, Helsinki University of Technology 1989

Heikkinen - Komonen Architects, 1985–
partner 2010–

Markku Puumala

b. 1964

Master of Science in Architectur
Helsinki University of Technolog
2005

Heikkinen – Komonen Architects
1996-
partner and CEO 2010–